METHODS OF
NONLINEAR ANALYSIS

Volume II

This is Volume 61-II in
MATHEMATICS IN SCIENCE AND ENGINEERING
A series of monographs and textbooks
Edited by RICHARD BELLMAN, *University of Southern California*

The complete listing of books in this series is available from the Publisher upon request.

METHODS OF NONLINEAR ANALYSIS

Richard Bellman

Departments of Mathematics,
Electrical Engineering, and Medicine
University of Southern California
Los Angeles, California

VOLUME II

1973

ACADEMIC PRESS New York and London

ACADEMIC PRESS, INC.
111 Fifth Avenue, New York, New York 10003

United Kingdom Edition published by
ACADEMIC PRESS, INC. (LONDON) LTD.
24/28 Oval Road, London NW1

LIBRARY OF CONGRESS CATALOG CARD NUMBER: 78-91424

PRINTED IN THE UNITED STATES OF AMERICA

To EMIL SELETZ

Surgeon, Sculptor, Humanitarian, and Friend

PREFACE

This is the second of two volumes written to introduce the reader to some of the theories and methods which enable us to penetrate carefully and timorously into the nonlinear domain. Fortunately, we have no choice: the only direction is forward. In Volume I we focused on the basic concepts of stability and variational analysis and the analytic techniques required for their use such as linear differential equations and matrix analysis.

In this volume we wish to describe some newer approaches such as duality techniques, differential inequalities, quasilinearization, dynamic programming, and invariant imbedding as well as some older methods which have become operational, and thus of greater analytic interest, as a consequence of the development of the digital computer: iteration, infinite systems of differential equations, and differential and integral quadrature. Using these theories we shall be able to present a more satisfactory explanation of various phenomena noted in Volume I, and indicate some promising new techniques.

A theme that runs through both volumes is that of closure, a closure from the standpoint of both formulations and algorithms. This is essential since we are interested in developing solution procedures which are feasible in terms of both contemporary analysis and computer technology. Indeed, it must be kept in mind that as far as can be seen into the future, routine computational techniques will not suffice to provide numerical solutions of equations of contemporary interest, nor will conventional formulations always be powerful enough. To a great degree, the more modern computers require more powerful mathematics for their effective use, as well as a liberal sprinkling of ingenuity; contemporary systems need new ideas.

Let us now briefly describe the contents of this volume.

In the chapter on duality, Chapter 9, we begin with the observation that a minimization process possesses the great advantage of furnishing immediate upper bounds. Thus, $\min_u J(u) \leqslant J(u_1)$ for any admissible function u_1. A crucial next step is to find an associated maximization process, say a functional $K(v)$, such that $\max_v K(v) = \min_u J(u)$. Then, since for any admissible v_1 we have

$$K(v_1) \leqslant \max_v K(v) = \min_u J(u) \leqslant J(u_1),$$

we can readily obtain upper as well as lower bounds. If these are close enough, the question of detailed solution of the original problem becomes academic.

Sometimes the physical background of the problem quickly furnishes a pair of functionals with the requisite property. Occasionally, geometric duality can be invoked, usually by way of the theory of convexity. In Chapter 9 we employ some simple analytic versions of these latter ideas which enable us to present significant results without getting deeply enmeshed in background material concerning convexity where a certain amount of mathematical sophistication is required.

Next we turn to a fruitful idea of Caplygin for obtaining approximate solution of a complex equation $T(u) = 0$. Let T_1 and T_2 be such that the solutions of $T_1(u) = 0$ and $T_2(u) = 0$ are reasonably simple to obtain, e.g., linear equations, or low order nonlinear equations. Suppose further that we have the inequalities

$$T_1(u) \leqslant T(u) \leqslant T_2(u) \qquad \text{for all } u.$$

Can the solutions of the simpler equations be used to bound the solution of the original equation $T(u) = 0$? Questions of this nature bring us in contact with differential inequalities and monotone operators, fields extensively cultivated by Collatz and his pupils. While in this area, we pay a brief visit to the second method of Lyapunov, a fundamental tool of stability theory.

It turns out that a systematic way of obtaining these bounding operators is furnished by the theory of quasilinearization. In many important cases we can write

$$T(u) = \max_{v} S(u, v),$$

and then show, by way of the above mentioned monotone operator theory, that the solution of $T(u) = 0$ may be written $u = \max_v W(v)$ where $W(v)$ is the solution of $S(u, v) = 0$. This is discussed, together with particularizations, in Chapter 11. Of these, the application to the Riccati equation is particularly interesting. The foregoing leads quite naturally to a method of successive approximations which yields the Newton–Raphson–Kantorovich procedure with the bonus of monotonicity in some cases.

As we shall see, equations of the form $\max_v S(u, v) = 0$ are characteristic of the theory of dynamic programming where u corresponds to the return of a multistage decision process and v to a policy. The observation that the calculus of variations, in both one-dimensional and multidimensional versions, may be viewed as a multistage decision process of deterministic type leads to new insights with many interesting analytic and computational consequences. Thus, for example, a simple and direct derivation of Hamilton–Jacobi theory is obtained and the Riccati equation is revealed as one of the fundamental equations of analysis. Furthermore, direct computational algorithms are provided for the minimization of functionals of the form $J(u) = \int_0^T g(u, u')\, dt$ which totally avoid the irksome two-point boundary value problems for the Euler equation furnished by the usual variational procedures, as well as questions of relative minima and stationary values. Analogous procedures can be used to furnish the minimum of such functionals as $\int_R (u_x^2 + u_y^2)\, dA$ over irregular regions.

The introduction of the concept of policy leads logically to that of approximation in policy space. The monotonicity expected intuitively from this interpretation of many classical equations is readily confirmed by the theory of monotone operators discussed in Chapter 10. Quasilinearization appears in this fashion as a natural outgrowth of the fundamental dynamic programming equation

$$f(p) = \max_q [g(p, q) + h(p, q) f(T(p, q))],$$

as was its motivation chronologically.

This novel type of functional equation can be used as a basis of a nonlinear semigroup theory, as well as a theory of stochastic and adaptive variational processes, directions which we do not pursue here. Let us merely note that it begins to appear that a fruitful extension of the classical linear theory, and probably the most fruitful one, is a nonlinear theory based upon the judicious use of the maximum operation, or, equivalently, imbedding in a decision or control process.

In Chapter 13 we turn to the theory of invariant imbedding, a theory devoted in part to the study of the different multistage processes which arise so naturally in mathematical physics. Once again, the basic role of the Riccati equation in analysis is clearly seen. The early work of Ambarzumian and Chandrasekhar in astrophysics played a crucial role in the development of the ideas of this theory. Conceptually, it can be regarded as a systematic method of treating the scientific world in terms of observables. Analytically, it can be viewed as an attempt to interpret processes as multistage processes and thus formulate them in terms of initial value problems (Volterra integral equations), in place of the boundary value problems (Fredholm integral equations), derived by classical imbedding.

Let us observe that, as is frequently the case, new and different formulations of classical processes often enable us to resolve some long-standing problem. In addition, a new foundation suggests many new questions. Generally, we want both as many different analytic formulations of a physical process as we can get, and identification of a particular equation with as many physical processes as possible.

Both dynamic programming and invariant imbedding are concerned with iteration, a fact which renders them so well adapted for the digital computer. In Chapter 14 we turn to the classical theory of iteration, briefly discussing its intimate connection with difference and differential equations and its rapidly expanding newer direction, the theory of branching processes.

Finally in Chapters 15 and 16 we return directly to matters of closure. In

Chapter 15 we consider the venerable technique of infinite systems of differential equations, together with truncation, to treat linear and nonlinear partial differential equations. In Chapter 16, we consider an alternate approach, namely the use of quadrature techniques, both the classical integral quadrature and the newer differential quadrature approach.

Some of the most important areas of contemporary nonlinear analysis have been regretfully omitted. Let us cite, in particular, the elegant and important area of fixed-point theorems, the field initiated by Birkhoff and Kellogg, the use of functional analysis, and the maximum principles of linear and nonlinear partial differential equations, so closely connected with much of the material we include.

Consistent with our desire to furnish an introduction rather than an Einführung we have avoided the temptation to be encyclopedic. Each chapter could easily be greatly expanded, and indeed into a book if desired. This has been done by the author and many others for all except Chapter 15. We have tried to amplify the text with extensive bibliographies and numerous exercises of greater or lesser importance and difficulty.

Secondly, we have devoted no time or attention to speak of to computational algorithms, neither those of deterministic nature such as finite difference algorithms or nonlinear programming, nor those of stochastic nature such as Monte Carlo methods.

Finally, it is a pleasure to acknowledge the help in many different ways of a number of friends and colleagues. Let us cite J. Casti, K. L. Cooke, T. J. Higgins, A. Lew, D. Tuey, and particularly Rebecca Karush who typed a plethora of drafts over the years.

RICHARD BELLMAN

Los Angeles, 1972

CONTENTS

Chapter 13. Invariant Imbedding

Chapter 14. The Theory of Iteration

Chapter 15. Infinite Systems of Ordinary Differential Equations and Truncation

Chapter 16. Integral and Differential Quadrature

CONTENTS OF VOLUME I

Chapter 9 UPPER AND LOWER BOUNDS VIA DUALITY

9.1. Introduction

In Volume I, we discussed the Bubnov–Galerkin and Rayleigh–Ritz methods, both procedures for transforming the task of solving a given equation into that of minimizing a specified functional. By means of a suitably chosen sequence of trial functions and algebraic operations we could obtain a decreasing sequence of upper bounds for the desired minimum value, and thus derive an approximation to the solution of the original equation.

The Rayleigh–Ritz method, powerful as it is, would become very much more potent if we possessed a corresponding technique for deriving lower bounds. One way to do this is to construct an associated maximization problem. In a number of important cases there is a straightforward way of generating the required comparison problem by utilization of the fundamental concept of duality.

What follows below is closely connected with a number of important areas of classical and modern mathematics, as indicated in the Bibliography and Comments at the end of the chapter. Fortunately, we can present a self-contained presentation of what is needed for our purposes here without being forced to introduce a large amount of extraneous and sophisticated material.

What follows is part of a much larger story concerning the contact transformations of Legendre, the mathematical theory of games, and the theory of convexity. The classical use of the Legendre transformation to obtain an associated functional is due to Friedrichs. The methods presented below appear to allow a greater flexibility in deriving associated variational problems, and therefore seem preferable at this stage. References to extensive work in the domains mentioned above will be found at the end of the chapter.

9.2. Guiding Idea

Let $J(u)$ be a functional whose minimum is desired as u varies over some prescribed set of functions. Suppose that, in some fashion or other, we can write $J(u)$ as the maximum of a second functional $K(u, v)$,

$$J(u) = \max_v K(u, v), \tag{9.2.1}$$

where v belongs to another set of functions. Then we can write, for any function v in the set,

$$J(u) \geqslant K(u, v), \qquad \min_u J(u) \geqslant \min_u K(u, v), \tag{9.2.2}$$

and thus, finally,

$$\min_u J(u) \geqslant \max_v \min_u K(u, v). \tag{9.2.3}$$

This provides a lower bound. In some cases, as we will indicate below, equality holds. This is to say,

$$\min_u \max_v K(u, v) = \max_v \min_u K(u, v). \tag{9.2.4}$$

This, however, is not of great importance for our immediate aims. What is more significant is that suitable trial functions for v in (9.2.2) yield the desired lower bounds for $\min_u J(u)$.

The operational success of this method depends on the following premises:

(a) We possess a systematic method for obtaining the functional $K(u, v)$.

(b) We can determine $H(v) = \min_u K(u, v)$. (9.2.5)

(c) We can readily apply the Rayleigh–Ritz method to the maximization of $H(v)$.

In a number of cases these desiderata are met.

9.3. A Simple Identity

In order to apply the general ideas sketched above to quadratic functionals, we will make use of the following simple scalar identity:

$$a^2 = \max_b (2ab - b^2). \tag{9.3.1}$$

The proof is immediate. We have

$$\begin{aligned} a^2 &= (b + a - b)^2 = b^2 + 2b(a - b) + (a - b)^2 \\ &\geqslant b^2 + 2b(a - b) = 2ab - b^2, \end{aligned} \tag{9.3.2}$$

with equality only for $b = a$.

9.4. Quadratic Functional: Scalar Case

Let us now consider the problem of minimizing the functional

$$J(u) = \int_0^T (u'^2 + \varphi(t)\, u^2)\, dt \tag{9.4.1}$$

over all functions u for which $u' \in L^2(0, T)$, and $u(0) = c$. We shall assume that $\varphi(t) > a > 0$ for $0 < t < T$, and that the function $\varphi(t)$ is differentiable. This is not absolutely required, but it simplifies the subsequent presentation to assume it.

The associated Euler equation is

$$u'' - \varphi(t)u = 0, \qquad u(0) = c, \quad u'(T) = 0. \tag{9.4.2}$$

The second condition is the free boundary condition which we may as well impose from the start.

Using the identity of Sec. 9.3, we can write

$$u'^2 = \max_v [2u'v - v^2]. \tag{9.4.3}$$

Hence,

$$J(u) = \max_v \left[\int_0^T [2u'v - v^2 + \varphi(t)\, u^2]\, dt\right]. \tag{9.4.4}$$

We have thus achieved the first step; the desired auxiliary functional is

$$K(u, v) = \int_0^T [2u'v - v^2 + \varphi(t)\, u^2]\, dt. \tag{9.4.5}$$

The next step is to determine

$$H(v) = \min_u K(u, v). \tag{9.4.6}$$

To do this, we eliminate the term $\int_0^T u'v\, dt$ by means of an integration by parts,

$$\int_0^T u'v\, dt = uv\Big]_0^T - \int_0^T uv'\, dt. \tag{9.4.7}$$

Since the function v which maximizes in (9.4.3) is equal to u' and since $u'(T) = 0$, we set $v(T) = 0$. Thus, (9.4.7) yields

$$\int_0^T u'v\,dt = -cv(0) - \int_0^T uv'\,dt. \tag{9.4.8}$$

Hence, we may write

$$K(u, v) = -2cv(0) + \int_0^T [\varphi(t)\,u^2 - 2uv' - v^2]\,dt. \tag{9.4.9}$$

The minimization over u is now readily carried out, namely,

$$\min_u K(u, v) = -2cv(0) + \int_0^T \left[-\frac{v'^2}{\varphi(t)} - v^2\right] dt. \tag{9.4.10}$$

The minimizing function is given by the relation

$$u = \frac{v'}{\varphi(t)}. \tag{9.4.11}$$

The determination of a lower bound for $\min_u J(u)$ thus depends upon our ability to maximize the companion functional

$$H(v) = -2cv(0) + \int_0^T \left[-\frac{v'^2}{\varphi(t)} - v^2\right] dt, \tag{9.4.12}$$

subject to the constraint $v(T) = 0$ and the condition that $v' \in L^2(0, T)$.

We see that the Euler equation associated with this functional is

$$\frac{d}{dt}\left(\frac{v'}{\varphi}\right) - v = 0, \tag{9.4.13}$$

subject to the boundary conditions

$$\left(\frac{v'}{\varphi}\right)_{t=0} = c, \qquad v(T) = 0. \tag{9.4.14}$$

Furthermore, the maximum value of H is given explicitly by

$$\max_v H(v) = -cv(0). \tag{9.4.15}$$

Approximate values for the maximum of $H(v)$, as also for the minimum value of $J(u)$, are to be calculated by means of the Rayleigh–Ritz procedure. We shall say more about this below.

9.5. $\min_u J = \max_v H$

Let us now turn to a demonstration of the equality of $\min_u J$ and $\max_v H$. The result can be obtained without calculation on the basis of general theorems based upon the fact that $K(u, v)$ is convex in u and concave in v. Let us, however, establish the equality in a straightforward fashion by noting that the minimum of $J(u)$, under the assumption concerning the positivity of φ, is determined by the solution of (9.4.2), and

$$\min_u J(u) = -cu'(0). \tag{9.5.1}$$

Writing (9.4.2) in the form

$$\frac{(u')'}{\varphi} - u = 0, \tag{9.5.2}$$

and differentiating, we see that u' satisfies the equation

$$\frac{d}{dt}\left(\frac{(u')'}{\varphi}\right) - u' = 0, \qquad u'(T) = 0, \quad \left(\frac{(u')'}{\varphi}\right)_{t=0} = c. \tag{9.5.3}$$

Hence, from the solution of (9.4.13) and (9.4.14), $v = u'$, as we suspected. Comparing (9.4.15) and (9.5.1), we see that we have established the desired equality.

9.6. The Functional $\int_0^T [u'^2 + g(u)]\, dt$

Let us now apply these methods to the problem of obtaining lower bounds for the minimum of the functional

$$J(u) = \int_0^T [u'^2 + g(u)]\, dt, \tag{9.6.1}$$

under the assumption that g is strictly convex. This, as we know (Sec. 7.23) guarantees that the minimum of $J(u)$ exists and is determined by the unique solution of

$$2u'' - g'(u) = 0, \qquad u(0) = c, \quad u'(T) = 0. \tag{9.6.2}$$

Write, as before,

$$J(u) = \max_v K(u, v), \tag{9.6.3}$$

where

$$K(u, v) = \int_0^T [2u'v - v^2 + g(u)]\, dt$$
$$= -2cv(0) + \int_0^T [-2uv' + g(u) - v^2]\, dt. \qquad (9.6.4)$$

Then

$$\min_u J(u) \geqslant \min_u K(u, v) \qquad (9.6.5)$$

for all v, and

$$\min_u K(u, v) = H(v) = -2cv(0) + \int_0^T [G(v') - v^2]\, dt, \qquad (9.6.6)$$

where we define the function G by means of the relation

$$G(w) = \min_u [g(u) - 2uw]. \qquad (9.6.7)$$

Since, by assumption, $g(u)$ is convex, we have

$$g'(u) - 2w = 0 \qquad (9.6.8)$$

as the equation determining the minimizing value of u, and

$$G'(w) = (g'(u) - 2w)\frac{du}{dw} - 2u = -2u. \qquad (9.6.9)$$

The function $G(w)$ is thus seen to be concave in w.

The Euler equation associated with $H(v)$ is

$$(G'(v'))' + 2v = 0, \qquad (9.6.10)$$

with the conditions

$$v(T) = 0, \qquad G'(v')_{t=0} = -2c. \qquad (9.6.11)$$

The concavity of G guarantees the uniqueness of solution of (9.6.10).

It follows, comparing (9.6.10), (9.6.11) with (9.6.8), (9.6.9) and (9.6.12), that $v = u'$ is a solution of (9.6.10), and therefore the solution. From this it follows that $J(u)$ has the same minimum value as the maximum of $H(v)$.

Exercises

1. What happens if we ignore the fact that $v(T) = 0$ and write

$$H(v) = 2u(T)\, v(T) - 2cv(0) + \int_0^T [\varphi u^2 - 2uv' - v^2]\, dt\ ?$$

2. Carry through the process for the case where the conditions are

   (a) $u(0) = c_1$, $u(T) = c_2$, or (b) $u(0) = c_1$, $u'(0) + bu(0) = c_2$.

3. Obtain the corresponding results for the functional.

$$J(u) = \int_0^T [p(t)\, u'^2 + q(t)\, u^2]\, dt, \qquad \text{where} \quad p, q > 0.$$

4. Consider the sequence $\{u_n\}$ generated by the successive minimization of

$$G(u_n, u_{n-1}) = \int_0^T [u_n'^2 + \varphi(u_{n-1}) + (u_n - u_{n-1})\, \varphi'(u_{n-1})]\, dt,$$

   $n = 1, 2, \ldots$, where u_0 is given, subject to $u_n(0) = c$, and consider the analogous sequence formed from $H(v)$. Discuss the convergence of these sequences.

5. Does a duality equality hold for each n? Can one establish the foregoing result in this way?

6. Consider the multidimensional case.

7. Consider the analogous problem for the case where

$$G(u_n, u_{n-1}) = \int_0^T \Big[u_n'^2 + \varphi(u_{n-1}) + (u_n - u_{n-1})\, \varphi'(u_{n-1}) + \frac{(u_n - u_{n-1})^2}{2}\, \varphi''(u_{n-1})\Big]\, dt.$$

9.7. Geometric Aspects

The reader may wonder where the identity of Sec. 9.3 originated. Although it can be viewed solely as an ingenious analytic device, it is worthwhile to note that the relation has a very simple geometric interpretation. Consider the graph of the parabola $y = x^2$ in Fig. 9.1, and the tangent to the curve at the point (u, u^2). Its equation has the form

$$y - u^2 = 2u(x - u), \qquad y = 2ux - u^2. \tag{9.7.1}$$

Since the parabola is a convex curve, the curve lies completely above the tangent except at the point of contact. Hence,

$$x^2 \geqslant 2ux - u^2 \tag{9.7.2}$$

with equality at $u = x$.

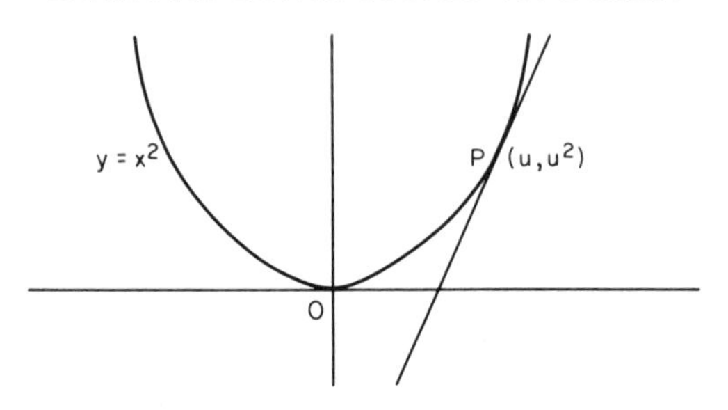

Figure 9.1

This argument extends to an arbitrary curve $y = g(x)$, where $g(x)$ is strictly convex, i.e., $g''(x) > 0$. We see that

$$g(x) \geqslant g(u) + (x - u)\, g'(u) \tag{9.7.3}$$

with equality only at $u = x$. Hence, we may write

$$g(x) = \max_u [g(u) + (x - u)\, g'(u)]. \tag{9.7.4}$$

We will use this result below.

9.8. Multidimensional Case

The same ideas are easily extended to the multidimensional case. Consider the functional

$$J(x) = \int_0^T [(x', x') + (x, A(t)x)]\, dt, \tag{9.8.1}$$

where $A(t)$ is positive definite for $0 \leqslant t \leqslant T$. We wish to minimize $J(x)$ over all vectors x for which $x(0) = c$ and $x' \in L^2(0, T)$.

The extension of the identity of Sec. 9.3 is the relation

$$(y, y) = \max_z [2(y, z) - (z, z)]. \tag{9.8.2}$$

Hence,

$$J(x) = \max \left[\int_0^T [2(x', z) - (z, z) + (x, A(t)x)]\, dt\right]. \tag{9.8.3}$$

Thus, integrating by parts as before and taking cognizance of the condition $x'(T) = 0$, we have

$$J(x) \geqslant \left[-2(c, z(0)) + \int_0^T [(x, A(t)x) - 2(x, z') - (z, z)]\, dt\right] \tag{9.8.4}$$

for all $z' \in L^2(0, T)$. Hence,

$$\min_x J(x) \geqslant \min_x[\cdots]. \tag{9.8.5}$$

The minimization over x is easily performed. The minimum of $(x, A(t)x) - 2(x, z')$ is attained for $x = A(t)^{-1}z'$, yielding the relation

$$\min_x J(x) \geqslant H(z) = -2(c, z(0)) + \int_0^T [-(z', A(t)^{-1} z') - (z, z)]\, dt. \tag{9.8.6}$$

(See Sec. 9.12, Exercise 1.) The maximum of the functional $H(z)$ over $z' \in L^2(0, T)$ is readily obtained. We leave as a series of exercises for the reader the task of showing that actually

$$\min_x J(x) = \max_z H(z). \tag{9.8.1}$$

Exercise

1. Obtain in the same way a maximization problem equivalent to the problem of minimizing $J(x) = \int_0^T [(x', x') + g(x)]\, dt$, where $g(x)$ is a strictly convex function of x.

9.9. The Rayleigh–Ritz Method

Consider the problem of minimizing

$$J(x) = \int_0^T [(x', x') + (x, A(t)x)]\, dt, \tag{9.9.1}$$

subject to $x(0) = c$, $x' \in L^2(0, T)$, and the associated functional

$$H(z) = -2(c, z(0)) + \int_0^T [-(z', A(t)^{-1} z') - (z, z)]\, dt. \tag{9.9.2}$$

We can, of course, apply the Rayleigh–Ritz method independently to both functionals. Since, however, we know that $z = x'$ for the pair of extremals, it is perhaps more reasonable to preserve this relation with the trial functions. Thus, if we set

$$x_N = \sum_{k=1}^{N} a_k y_k(t), \tag{9.9.3}$$

where

$$\sum_{k=1}^{N} a_k y_k(0) = c, \qquad \sum_{k=1}^{N} a_k y_k'(T) = 0, \tag{9.9.4}$$

it is appropriate to use

$$z_N = \sum_{k=1}^{N} a_k y_k'(t) \tag{9.9.5}$$

as a trial function for $H(z)$.

9.10. Alternative Approach

Consider another approach to the problem of minimizing the quadratic functional

$$J(u) = \int_0^T (u'^2 + \varphi(t)\, u^2)\, dt \tag{9.10.1}$$

over all $u(t)$ such that $u' \in L^2(0, T)$ and for which $u(0) = c$. As we know, the Euler equation is

$$u'' - \varphi(t)u = 0, \qquad u(0) = c, \quad u'(T) = 0, \tag{9.10.2}$$

and the minimum value of $J(u)$ is $-u'(0)c$, determining the missing initial condition.

To obtain a lower bound, we once again employ the identity

$$u^2 = \max_v (2uv - v^2), \tag{9.10.3}$$

with the maximum attained at $v = u$. Let us now assume that $\varphi(t)$ is positive. We can then use the relation in connection with the term $\varphi(t)u^2$. We have

$$\int_0^T (u'^2 + \varphi(t)\, u^2)\, dt = \max_{v(t)} \int_0^T (u'^2 + \varphi(t)(2uv - v^2))\, dt. \tag{9.10.4}$$

Hence,

$$\min_u \int_0^T (u'^2 + \varphi(t)\, u^2)\, dt = \min_u \max_v \int_0^T (u'^2 + \varphi(t)(2uv - v^2))\, dt. \tag{9.10.5}$$

The minimization with respect to u of the functional

$$L(u) = \int_0^T (u'^2 + 2\varphi(t)\, uv)\, dt, \tag{9.10.6}$$

subject to $u(0) = c$, is readily carried out. The Euler equation is

$$u'' - \varphi(t)v = 0, \qquad u(0) = c, \quad u'(T) = 0, \tag{9.10.7}$$

yielding the explicit solution

$$u = c + \int_0^T k(t, t_1)\, \varphi(t_1)\, v(t_1)\, dt_1 , \tag{9.10.8}$$

where k is the Green's function associated with the linear equation $u'' = 0$ and the boundary conditions $u(0) = 0$, $u'(T) = 0$. The value of the minimum is readily seen to be a quadratic functional of v.

An explicit calculation of the maximum of this functional with respect to v shows that the lower bound is attained and that once again min–max = max–min.

Exercises

1. Extend the foregoing method to obtain lower bounds for the minimum of $J(u) = \int_0^T [u'^2 + g(u)]\, dt$, $u(0) = c$, where it is assumed that $g(u)$ is strictly convex.

2. Consider the problem of minimizing the quadratic functional

$$J_n(u) = \int_0^T [u'^2 + g(u_n) + g'(u_n)(u - u_n) + g''(u_n)(u - u_n)^2/2]\, dt,$$

where $g(u)$ is strictly convex and u_n is an nth approximation. Take $u_0 = c$. The minimum of $J_n(u)$ is furnished by u_{n+1}. Write down the Euler equation and the Euler equation for the associated maximization problem.

3. Under what conditions on $g(u)$ is the sequence of successive approximations, $\{\min J_n(u)\}$ monotone in n.

4. If $g(u)$ is convex show that

$$\min_u J(u) \geqslant \min_u \left[\int_0^T (u'^2 + g(u_n)(u - u_n))\right] dt$$

for any function u_n.

5. Extend the foregoing to the vector case where

$$J(x) = \int_0^T [(x', x') + (x, A(t)x)]\, dt$$

and $A(t)$ is positive definite.

6. Similarly, consider the function

$$J(x) = \int_0^T [(x', x') + g(x)]\, dt,$$

where $g(x)$ is a convex function of x.

7. Extend both of the foregoing methods to functionals of the form

$$J(u) = \int_R [u_x^2 + u_y^2 + e^u]\, dA,$$

$u = g$, on B the boundary of R.

8. Show that

$$\min_u \int_0^1 (u'^2 + \varphi(u))\, dt = \max_v \Big[\tfrac{1}{4} \int_0^1 \int_0^1 k(s, t)\, \varphi'[v(s)]\, \varphi'[v(t)]\, ds\, dt + \int_0^1 [(c - v)\, \varphi'(v) + \varphi(v)]\, dt\Big],$$

where $k(s, t)$ is the Green's function associated with $w'' = f(t)$, $w(0) = w'(1) = 0$, and the minimization with respect to u is over functions satisfying $u(0) = c$.

9. Apply the foregoing to obtain an approximate solution of

$$u'' + u + \epsilon u^3 = 0, \qquad u(0) = 1, \qquad u'(1) = 0, \qquad \text{for} \quad 0 < \epsilon < 1.$$

9.11. $J(u) = \int_0^T [u'^2 + \varphi(t)u^2]\, dt$; General $\varphi(t)$

In the foregoing sections, we have insisted upon the fact that the expression $u'^2 + \varphi(t)u^2$ be positive definite as a function of u' and u. Analyzing what is actually required, we see that it is sufficient to assume that the functional $J(u)$ is positive definite, the minimum requirement for the existence of a minimum. Let us then consider

$$J(u) = \int_0^T [u'^2 + \varphi(t)\, u^2]\, dt, \tag{9.11.1}$$

where the requirement is now that $J(u)$ be positive definite for all u satisfying $u(0) = 0$, $u'(T) = 0$. This is certainly the case if T is sufficiently small. Then

$$J(u) \geqslant \lambda_1 \int_0^T u^2\, dt \tag{9.11.2}$$

for some $\lambda_1 > 0$ for all u in the foregoing class; see Chapter 8, Volume I.

Write

$$J(u) = \int_0^T [(u'^2 + (\varphi(t) - a)\,u^2) + au^2]\,dt, \tag{9.11.3}$$

where a is positive, and chosen so that

$$J_a(u) = \int_0^T [u'^2 + (\varphi(t) - a)\,u^2]\,dt \tag{9.11.4}$$

is positive definite. This quantity a exists by virtue of (9.11.2); for example, we may take $a = \lambda_1/2$.

Since $J_a(u)$ is positive definite, we have, expanding the integrand,

$$\begin{aligned} J_a(u) &= J_a(v + u - v) \\ &= J_a(v) + 2\int_0^T [v'(u' - v') + (\varphi(t) - a)\,v(u - v)]\,dt + J_a(u - v). \end{aligned} \tag{9.11.5}$$

Hence,

$$J_a(u) \geqslant J_a(v) + 2\int_0^T [v'(u' - v') + (\varphi(t) - a)\,v(u - v)]\,dt \tag{9.11.6}$$

for all v with equality if and only if $u = v$. Thus, collecting terms we may write

$$J_a(u) = \max_v \left[2\int_0^T (v'u' + (\varphi(t) - a)\,vu)\,dt - \int_0^T (v'^2 + (\varphi(t) - a)\,v^2)\,dt\right]. \tag{9.11.7}$$

Proceeding as before, we can write

$$\begin{aligned} \min_u J(u) &= \min_u \left[J_a(u) + a\int_0^T u^2\,dt\right] = \min_u \left[\max_v[\cdots] + a\int_0^T u^2\,dt\right] \\ &\geqslant \max_v \min_u \left[[\cdots] + a\int_0^T u^2\,dt\right] = \max_v H(v). \end{aligned} \tag{9.11.8}$$

A direct calculation shows that the two extremes are actually equal.

9.12. Geometric Aspects

It is interesting to consider the geometric significance of what we have been doing. The functional $J(u)$ can be considered the square of a distance from some fixed point P to a point Q in a set S, where u is an element of S. From the way we have drawn Fig. 9.2, it is clear we need only consider boundary points. Here d represents the minimum distance. This minimum distance, however, can also be considered to be the

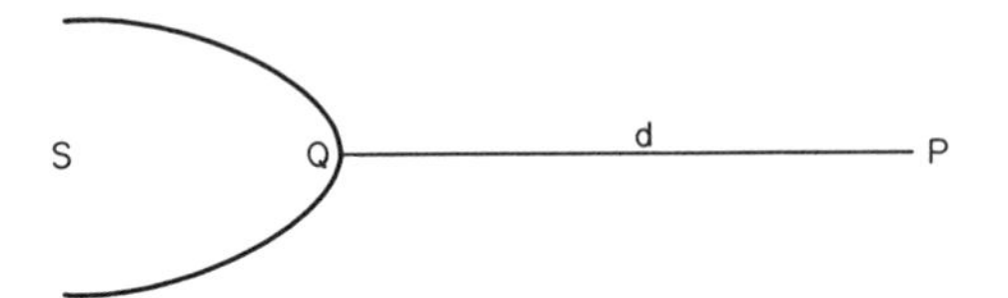

Figure 9.2

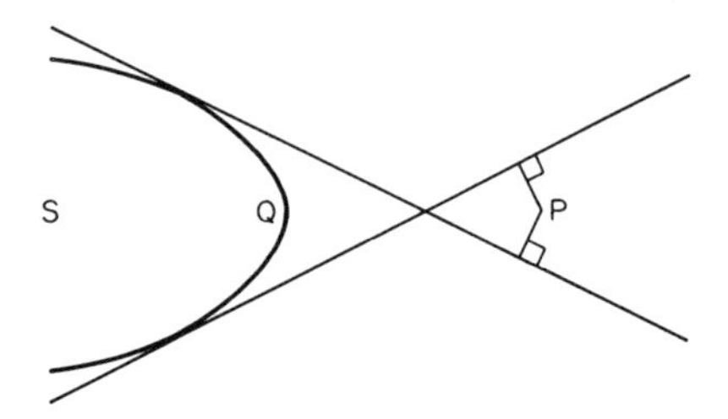

Figure 9.3

maximum of distances from P to the tangent lines at the points of the boundary of S, as indicated in Fig. 9.3.

We seen then how a minimization problem can be made equivalent to a maximization problem. The fact that the boundary of S can be described either as a locus of points or as an envelope of tangents is a manifestation of the fundamental duality concept that dominates analysis, algebra, and geometry.

Exercise

1. Carry out the details for the minimum distance from a point to a circle, an ellipse, and a sphere.

Miscellaneous Exercises

1. Suppose that A is a positive definite matrix. Show that $Ax = y$ is the equation obtained from the minimization of $(x, Ax) - 2(x, y)$, and that $-(y, A^{-1}y) = \min_x [(x, y) - 2(x, Ax)]$.

2. Use a trial function $x^{(M)} = \sum_{k=1}^{M} c_k x_k$ to determine an approximation to the minimum in the foregoing problem. Write

$$(y, B_M y) = \min_{c_k} \left[\sum_{k,l=1}^{M} c_k c_l (x_k , Ax_l) - 2 \sum_{k=1}^{M} c_k (x_k , y) \right].$$

Show that $B_1 \geqslant B_2 \geqslant \cdots \geqslant B_M \geqslant A^{-1}$ where the relation $B_1 \geqslant B_2$ for symmetric matrices means, as before, that $B_1 - B_2$ is nonnegative definite.

3. Specializing the value of y, show that this implies $b_{ii}^{(M)} \geqslant b_{ii}^{(M+1)}$, where $b_{ij}^{(M)}$ denotes the ijth element of B_M. Show also that

$$(b_{ij}^{(M)} - b_{ij}^{(M+1)})^2 \leqslant (b_{ii}^{(M)} - b_{ii}^{(M+1)})(b_{jj}^{(M)} - b_{jj}^{(M+1)}).$$

4. Can one use small values of M plus extrapolation in M to estimate both the elements of A^{-1} and the minimizing x?

5. Let A be a real, not necessarily symmetric, matrix and B be a positive definite matrix. Show that the minimization over x and y of

$$J(x, y) = (x, Bx) + 2(x, Ay) + (y, By) - 2(a, x) - 2(b, y)$$

yields the equations $Bx + Ay = a$, $A'x + By = b$.

6. Let A be complex, $A = B + iC$, where B and C are real and symmetric. Show that if C is positive definite, then

$$\min_x \max_y J(x, y) = \max_y \min_x J(x, y),$$

where

$$J(x, y) = (x, Cx) + 2(x, By) - (y, Cy) - 2(b, x) - 2(a, y).$$

7. Consider the minimization of $J(u) = \int_0^T [u'^2 + \varphi(t)u^2]\, dt$ over polygonal functions with M vertices, satisfying $u(0) = c$. Let J_M denote this minimum. Show that $J_1 \geqslant J_2 \geqslant \cdots$. Can we use extrapolation techniques?

8. Consider the equation

$$u'' - x^{-1/2}u^{3/2} = 0, \qquad u(0) = 1, \quad u(\infty) = 0,$$

variously called the Emden–Fowler or Fermi–Thomas equation, depending upon whether one is working in astrophysics or nuclear physics. Show that it is the Euler equation associated with the functional

$$J(u) = \int_0^\infty (u'^2 + 4x^{-1/2}u^{5/2}/5)\, dx.$$

9. Consider the new functional

$$I(v) = -2v(0) - \int_0^\infty (v^2 + 6x^{1/2}(v')^{5/3}/5)\, dx.$$

Show that $I(v) \leqslant J(u)$ for all v.

10. Show that

$$\frac{5}{2}\left(\frac{2}{7}\right)^{4/3}(-v(0))^{7/3}\left(\int_0^\infty v^2\,dx\right)-\frac{1}{3}\left(\int_0^\infty x^{1/3}(v')^{5/3}\,dx\right)^{-1}$$
$$\leqslant J(u)\leqslant\frac{7}{2}\left(\frac{2}{5}\right)^{2/3}u(0)^{\,7/3}\left(\int_0^\infty u'^2\,dx\right)^{1/3}\left(\int_0^\infty x^{-1/3}u^{5/2}\,dx\right)^{2/3}$$

for any functions u and v for which the integrals exist and such that $u \geqslant 0$, $v' \geqslant 0$.

11. Use the trial function $u(x) = (1 + x^{1/\alpha})^{-\beta}$, $\infty > x > 0$, and show by appropriate choice of α and β that $J(u) \leqslant 1.5883$. Take

$$\begin{aligned} v(x) &= -(1 - x^{1/\alpha})^{\beta}, && 0 < x \leqslant 1, \\ &= 0, && x > 1, \end{aligned}$$

and show that $J(u) \geqslant 1.5865$. For the foregoing, see

T. Ikebe and T. Kato, "Application of Variational Method to the Thomas-Fermi Equation," *J. Phys. Soc. Japan*, Vol. 12, 1957, pp. 201–203,

where other references may be found. For an extension to some three-dimensional cases, see

Y. Tanabe, "On the Variational Problem Connected with the Thomas–Fermi Method," *J. Phys. Soc. Japan*, Vol. 12, 1957, p. 974.

For a systematic discussion of the properties of the solutions of $u'' \pm x^m u^n = 0$, $u'' - e^{\lambda x} u^n = 0$, see

R. Bellman, *Stability Theory of Differential Equations*, Dover, New York, 1970,

where references to earlier and more extensive work of Emden and Fowler may be found.

12. Consider the problem of minimizing the quadratic form

$$J_N(x) = \sum_{i=1}^{N} [\lambda_i x_i^2 + \mu_i (x_{i-1} - x_i)^2],$$

where λ_i, $\mu_i > 0$ and $x_0 = c$. Find an equivalent maximization problem.

13. Consider the function $g(v)$ associated with $f(u)$, $-\infty < u < \infty$ by means of the relation

$$g(v) = \max_u [uv - f(u)] \qquad \text{(the Fenchel transform).}$$

Under what conditions on $f(u)$ may we write

$$f(u) = \min_v [uv - g(v)] \ ?$$

Obtain the associated pairs of functions for the cases $f(u) = u^a$, $a > 1, f(u) = e^{bu}$, $b > 0$.

14. Consider the function $h(v)$ associated with $f(u)$ by means of the relation

$$h(v) = \max_{u \geq 0} [uv - f(u)] \qquad \text{(the Maximum transform).}$$

Under what conditions on $f(u)$ may we write

$$f(u) = \min_{v \geq 0} [uv - h(v)] \ ?$$

15. Consider the convolution

$$k(u) = \max_{0 \leq v \leq u} [f(v)\, g(u - v)] = f \circ g.$$

Let

$$M(f) = \max_{u \geq 0} [e^{-uv} f(u)] = F(v).$$

Show that

$$M(f \circ g) = M(f)\, M(g)$$

under suitable assumptions. For the foregoing, see

R. Bellman and W. Karush, "On a New Functional Transform in Analysis: The Maximum Transform," *Bull. Amer. Math. Soc.*, Vol. 67, 1961, pp. 501–503.

See also

R. Bellman and W. Karush, "Mathematical Programming and the Maximum Transform," *SIAM J. Appl. Math.*, Vol. 10, 1962, pp. 550–567.

R. Bellman and W. Karush, "On the Maximum Transform and Semigroups of Transformations," *Bull. Amer. Math. Soc.*, Vol. 68, 1962, pp. 516–518.

The maximum transform is related to the Fenchel transform in the same manner that the Laplace transform is connected with the Fourier transform, See

P. Whittle, "The Deterministic Stochastic Transition in Control Processes and the Use of Maximum and Integral Transforms," *Proc. Berkeley Symp. Math. Statist. Probabil., 5th*, Vol. 3, Phys. Sci. and Eng. (L. Lecan and J. Neyman, eds.), Univ. of California Press, Berkeley, California, 1967, pp. 217–228.

16. Show that we can write

$$\max_{t\geqslant 0}\left(st-\int_0^t M(t_1)\,dt_1\right)=\int_0^s M^{-1}(t_1)\,dt_1\,,$$

under an appropriate condition on $M(t)$. For a use of this result to derive interesting classes of inequalities, see

R. Redheffer, "Recurrent Inequalities," *Proc. London Math. Soc.*, Vol. 17, 1967, pp. 683–699.

For other applications of quasilinearization of this nature in inequality theory, see

E. F. Beckenbach and R. Bellman, *Inequalities*, Springer-Verlag, Berlin, 1961.

17. Consider the problem of minimizing

$$J_1(v)=\int_0^T |\,1-u\,|\,dt$$

with respect to v where $u'=-u+v$, $u(0)=1$, $0\leqslant v\leqslant m>1$, $\int_0^T v\,dt\leqslant a<T$. Write

$$|\,1-u\,|=\max_{\varphi}\varphi(1-u)$$

and thus

$$\min_v J_1(v)=\min_v\max_{\varphi}\int_0^T \varphi(1-u)\,dt.$$

Determine

$$\max_{\varphi}\min_v\int_0^T \varphi(1-u)\,dt$$

and show that $\max_{\varphi}\min_v=\min_v\max_{\varphi}$.

18. Consider the problem of minimizing

$$J_2(v) = \max_{0 \leqslant t \leqslant T} | 1 - u |$$

subject to the same constraints. Show that

$$J_2(v) = \min_v \max_{|\varphi| \leqslant 1} \varphi(1 - u) = \max_{|\varphi| \leqslant 1} \min_v \varphi(1 - u)$$

and thus determine the minimizing v. See

R. Bellman, I. Glicksberg, and O. Gross, "Some Unconventional Problems in the Calculus of Variations," *Bull. Amer. Math. Soc.*, Vol. 7, 1956, pp. 81–94,

R. Bellman, I. Glicksberg, and O. Gross, *Some Aspects of the Mathematical Theory of Control Processes*, R–313, RAND Corp., 1958, pp. 203–210. Russian Translation, Foreign Literature Press, Moscow, 1962.

See also

C. D. Johnson, "Optimal Control with Chebyshev Minimax Performance Index," *J. Basic Eng.*, 1967, pp. 251–262.

C. D. Johnson, "Singular Solutions in Problems of Optimal Control," *Advan. Control Syst.*, Vol. 2, 1965, pp. 209–267.

For extensions of the methods of the foregoing exercises, applications and other techniques, see

Gabasov, R., and F. M. Krillova, "Optimization of Linear Systems by the Methods of Functional Analysis," *J. Optimization Theory Appl.*, Vol. 8, 1971, pp. 77–99.

Bibliography and Comments

A further discussion and extensive references may be found in Chapter 9 of

Bellman, R., *Introduction to the Mathematical Theory of Control Processes*, Vol. 2, *Nonlinear Processes*, Academic Press, New York, 1971.

Some additional references of interest are

Gilleman, A., "Dual Formulations of Optimal Control Problems," *J. Optimization Theory Appl.*, Vol. 8, 1971, pp. 237–255.

Heins, W., and S. K. Mitter, "Conjugate Convex Functions, Duality and Optimal Control Problems I: Systems Governed by Ordinary Differential Equations," *Information Sci.*, Vol. 2, 1970, pp. 211–243.

Duffin, R. J., *Duality Inequalities of Mathematics and Science, Nonlinear Programming*, Univ. of Wisconsin, 1970, pp. 401–423.

J. J. Moreau, "Convexity and Duality," in *Functional Analysis and Optimization*, E. R. Caianiello (ed.), Academic Press, New York, 1966.

A. Ioffe and V. Tikhomirov, "Duality of Convex Functions and Extremum Problems," *Russian Math. Surveys*, Vol. 23, 1968, pp. 53–124.

R. Rockafellar, "Conjugate Convex Functions in Optimal Control and the Calculus of Variations," *J. Math. Anal. Appl.*, 1970, pp. 174–222.

See

D. J. White, "Envelope Programming and a Minimax Theorem," *J. Math. Anal. Appl.*, Vol. 39, 1972.

for a discussion of the minimization of functions which can be written in the form $\max \phi(x)_{x \in R}$ where $\phi(x)$ is strictly concave and R is convex.

Chapter 10 CAPLYGIN'S METHOD AND DIFFERENTIAL INEQUALITIES

10.1. Introduction

One of the first traumatic facts that confronts the mathematician is that the great majority of equations that arise in scientific investigations are analytically intractable. The challenge then to the analyst is to obtain qualitative and quantitative information concerning the solution without the benefit of an explicit representation in terms of the familiar functions and operations of analysis. The ingenious idea of Caplygin was to replace the original differential equation by suitably chosen differential inequalities whose associated equations can readily be solved to yield upper and lower bounds for the desired solution.

To apply this useful general method requires a certain amount of the theory of positive or, more precisely, monotone operators, a theory which owes so much to the fundamental work of L. Collatz. A number of different results and particular methods of this contemporary theory will be presented and analyzed. The results are both simpler and more complete for ordinary differential equations. Nevertheless, as far as possible, we emphasize techniques which can be used with slight modification to handle the corresponding problems for partial differential equations since much of the contemporary theory of partial differential equations centers around this problem area. Extensive references will be found at the end of the chapter.

In the final part of the chapter, we turn to Lyapunov's second method for establishing stability, a method which can be considered to be a direct application of the theory of differential inequalities. This technique while quite different from that presented in Chapter 3, Volume I, to study perturbed equations, is far more versatile, but in return yields less information.

10.2. The Caplygin Method

To illustrate the basic concept which is quite straightforward, suppose that we are given the equation

$$\frac{du}{dt} = u + \epsilon g(u), \qquad u(0) = 1, \tag{10.2.1}$$

where $g(u)$ is a function which lies between -1 and 1 for all values of u, and ϵ is a parameter. For example, $g(u)$ might be $\sin u$. Consider the associated linear equations

$$\frac{du_1}{dt} = u_1 + \epsilon, \qquad \frac{du_2}{dt} = u_2 - \epsilon, \tag{10.2.2}$$

subject to the same initial conditions. Can we assert that

$$u_1 \leqslant u \leqslant u_2 \tag{10.2.3}$$

in some t-interval $0 \leqslant t \leqslant t_0$?

If general results of this nature hold, we possess a systematic procedure for obtaining bounds on the solution of nonlinear differential equations by considering the solution of certain associated linear differential equations, or, generally, equations of a more tractable nature.

10.3. The Equation $u' \leqslant a(t)u + f(t)$

Let us begin with the first-order linear differential inequality

$$\frac{du}{dt} \leqslant a(t)u + f(t), \qquad u(0) = c, \tag{10.3.1}$$

where the treatment is particularly simple. We start with the fact that (10.3.1) is equivalent to the equation

$$\frac{du}{dt} = a(t)u + f(t) - p(t), \qquad u(0) = c, \tag{10.3.2}$$

where $p(t)$ is some function restricted by the condition that $p(t) \geqslant 0$ for $t \geqslant 0$.

The question we wish to examine is whether or not we can derive an inequality connecting the class of solutions of (10.3.2), which is to say the set of functions satisfying (10.3.1), and the solution of the associated equation

$$\frac{dv}{dt} = a(t)v + f(t), \qquad v(0) = c. \tag{10.3.3}$$

Let us demonstrate

Theorem. *Let u be a function satisfying* (10.3.1) *in the interval* $[0, t_0]$. *Then $u(t) \leqslant v(t)$ in this interval, where v is the solution of* (10.3.3).

The proof follows readily from the explicit representation of the solution of (10.3.2),

$$\begin{aligned} u &= c \exp\left[\int_0^t a(t_1)\,dt_1\right] + \exp\left[\int_0^t a(t_1)\,dt_1\right] \\ &\quad \times \left\{\int_0^t \exp\left[-\int_0^{t_1} a(s)\,ds\right][f(t_1) - p(t_1)]\,dt_1\right\} \\ &= v - \exp\left[\int_0^t a(t_1)\,dt_1\right] \int_0^t \exp\left[-\int_0^{t_1} a(s)\,ds\right] p(t_1)\,dt_1 . \qquad (10.3.4) \end{aligned}$$

Since $p \geqslant 0$ by assumption, we see that $u \leqslant v$.

Exercises

1. Show that the fundamental lemma of Chapter 1, Volume I, is a particular case of the foregoing result. (The inequality

$$u \leqslant c + \int_0^t uv\,dt_1 ,$$

can be considered to be a differential inequality,

$$\frac{dw}{dt} \leqslant cv + vw, \qquad w(0) = 0,$$

where $w = \int_0^t uv\,dt_1$.)

2. Show that the solution of the matrix equation

$$X' = A(t)X + X(t)\,B(t) + F(t), \qquad X(0) = C,$$

may be written

$$X = Y_1 C Y_2 + \int_0^t Y_1(t_1)^{-1}\,Y_1(t)\,F(t_1)\,Y_2(t)\,Y_2(t_1)^{-1}\,dt_1$$

where

$$Y_2'(t) = A(t)\,Y_2(t), \quad Y_2(0) = I, \qquad Y_2'(t) = Y_2(t)\,B(t), \quad Y_1(0) = I.$$

3. Deduce appropriate monotonicity relations from this representation.

10.4. The Linear Differential Inequality $L(u) \leqslant f(t)$

Let us next consider the nth-order case where we have the inequality

$$L(u) = u^{(n)} + a_1(t)\, u^{(n-1)} + \cdots + a_n(t)u \leqslant f(t), \tag{10.4.1}$$

and the equality

$$L(v) = f(t). \tag{10.4.2}$$

Supposing again that

$$u^{(k)}(0) = v^{(k)}(0) = 0, \qquad k = 0, 1, \ldots, n-1, \tag{10.4.3}$$

when can we assert that $u \leqslant v$ for $0 \leqslant t \leqslant t_0$, where t_0 is determined by the linear operator L.

A necessary and sufficient condition for the determination of t_0 can readily be given in terms of the kernel function $k(t, t_1)$, used to express the solution of the inhomogeneous equation in (10.4.2), when the initial conditions are $v^{(k)}(0) = 0$, $k = 0, 1, \ldots, n-1$, namely,

$$v = \int_0^t k(t, t_1)\, f(t_1)\, dt_1 . \tag{10.4.4}$$

Clearly the condition is $k(t, t_1) \geqslant 0$ for $0 \leqslant t_1 \leqslant t \leqslant t_0$. For the first-order equation discussed in Sec. 10.3, this condition is readily verified for any $t_0 \geqslant 0$. In general, the discussion is more complicated as might be imagined, since we do not possess any simple analytic expression for the solution of (10.4.2) for $n \geqslant 2$. Let us now examine a number of techniques which can be used to establish the nonnegativity of $k(t, t_1)$.

10.5. Elementary Approach

Let us begin with the first difficult case $n = 2$ using a slight extension of a result due to Caplygin himself. The kind of elementary reasoning used can often be applied to more complex functional equations.

Theorem. *If we have the relations*

$$\begin{aligned} &\text{(a)} \quad u'' + p(t)\, u' - q(t)u > 0, && t \geqslant 0, \\ &\text{(b)} \quad v'' + p(t)\, v' - q(t)v = 0, && t \geqslant 0, \\ &\text{(c)} \quad q(t) \geqslant 0, && t \geqslant 0, \\ &\text{(d)} \quad u(0) = v(0), \qquad u'(0) = v'(0), && \end{aligned} \tag{10.5.1}$$

then $u > v$ for $t > 0$.

Proof. Subtracting the equation in (10.5.1b) from the inequality in (10.5.1a), we have the relation

$$w'' + p(t)\,w' - q(t)w > 0, \tag{10.5.2}$$

where $w = u - v$, with $w(0) = w'(0) = 0$. Since this shows that $w''(0) > 0$, it follows that $w > 0$ in some initial interval $[0, t_0]$. Suppose that w becomes zero at some subsequent point t_1, and let t_1 be the first such point, as indicated in Fig. 10.1. Then w has a local maximum at some intermediate point t_2. At this point we have, from (10.5.2),

$$w'' + 0 - q(t_2)w > 0, \tag{10.5.3}$$

whence $w'' > 0$. This contradicts the assumption that t_2 provides a local maximum.

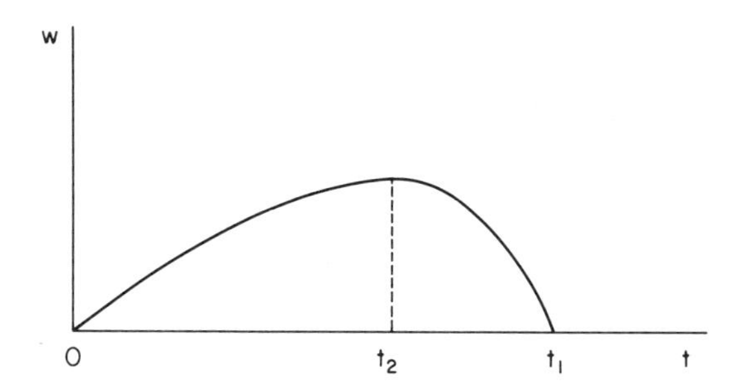

Figure 10.1

Exercises

1. Modify the proof to cover the case where (10.5.1a) contains a greater than or equal to sign.
2. Show that it is sufficient to establish the result for $p = 0$ to establish it for general p.

10.6. An Integral Identity

Caplygin's original proof depended upon the important identity

$$\int_0^{t_1} w(w'' - q(t)w)\,dt = [ww']_0^{t_1} - \int_0^{t_1} (w'^2 + q(t)\,w^2)\,dt, \tag{10.6.1}$$

used extensively in Chapter 8, Volume I. As noted above, it is sufficient to prove the theorem of Sec. 10.5 for the case where $p = 0$. From (10.6.1), we have

$$\int_0^{t_1} w(w'' - q(t)w)\, dt = - \int_0^{t_1} (w'^2 + q(t)\, w^2)\, dt < 0, \qquad (10.6.2)$$

by virtue of our assumption concerning q, if w is not identically zero. This contradicts the assumptions $w \geqslant 0$, $w'' - q(t)w \geqslant 0$ in $[0, t_1]$. Hence, the point t_1 does not exist.

Exercises

1. The problem of minimizing $J(u) = \int_0^T g(u, u')\, dt$ leads to the Euler equation $g_u - (d/dt)\, g_{u'} = 0 = E(u)$. We have the identity

$$0 = \int_0^T uE(u)\, dt = \int_0^T [ug_u + u'g_{u'}]\, dt = k \int_0^T g\, dt$$

if g is homogenous of degree k and the integrated terms vanish at the end points. What conclusions can we draw from this identity?

2. We have the matrix identity

$$\int_0^T X(X'' - A(t)X)\, dt = XX'\Big]_0^T - \int_0^T (X'^2 + XA(t)X)\, dt.$$

What conclusions can we obtain from this identity?

10.7. Strengthening of Previous Result

Examining the previous proof, we see that the condition $q(t) \geqslant 0$ was used only to ensure that the integral $\int_0^{t_1} (u'^2 + qu^2)\, dt$ is positive. If q is a function of more general type such that this integral is positive for $0 \leqslant t_1 \leqslant a$, then we can assert that the conclusion of the theorem of Sec. 10.5 is valid for $0 \leqslant t \leqslant a$. This is the positive definiteness condition that we employed in Chapter 8, Volume I. As we know, this is a characteristic value condition. Hence, we see the intimate connection between positivity and Sturm–Liouville theory.

10.8. Factorization of the Operator

Let us now introduce a useful factorization of the linear differential operator

$$L_2 = \frac{d^2}{dt^2} + p(t) \frac{d}{dt} + q(t). \qquad (10.8.1)$$

Let u_1, u_2 be two linearly independent solutions of $L_2(u) = 0$, ensuring that the Wronskian

$$W(u_1, u_2) = \begin{vmatrix} u_1 & u_1' \\ u_2 & u_2' \end{vmatrix} \neq 0 \tag{10.8.2}$$

(which means that we can take it positive), and write

$$W(u_1, u_2, u) = \begin{vmatrix} u_1 & u_1' & u_1'' \\ u_2 & u_2' & u_2'' \\ u & u' & u'' \end{vmatrix}. \tag{10.8.3}$$

Then we assert that

$$\begin{aligned} \frac{W(u_1, u_2, u)}{W(u_1, u_2)} &= \frac{W(u_1, u_2)}{u_1} \frac{d}{dt}\left(\frac{W(u_1, u)}{W(u_1, u_2)}\right) \\ &= \frac{W(u_1, u_2)}{u_1} \frac{d}{dt}\left(\frac{u_1^2}{W(u_1, u_2)} \frac{d}{dt}\left(\frac{u}{u_1}\right)\right). \end{aligned} \tag{10.8.4}$$

In Sec. 10.19, we present a proof of this important result.

Consider the determinantal equation $W(u_1, u_2, u) = 0$. Two solutions of this linear differential equation are clearly $u = u_1$, $u = u_2$. Hence,

$$\frac{W(u_1, u_2, u)}{W(u_1, u_2)} = u'' + p(t)\, u' + q(t)u. \tag{10.8.5}$$

Finally, we have the desired result.

Theorem. *If $u_1(t) > 0$ for $0 < t < t_0$, then in this interval*

$$u'' + p(t)\, u' + q(t)u = \frac{W(u_1, u_2)}{u_1} \frac{d}{dt}\left(\frac{u_1^2}{W(u_1, u_2)} \frac{d}{dt}\left(\frac{u}{u_1}\right)\right). \tag{10.8.6}$$

10.9. Alternate Proof of Monotonicity

Using the foregoing identity, we can readily establish the result of Sec. 10.5 under an apparently different condition.

Theorem. *Let the equation*

$$w'' + p(t)\, w' + q(t)w = 0 \tag{10.9.1}$$

possess a solution u_1 which is positive in $[0, t_0]$. Then in this interval the inequality

$$u'' + p(t)\, u' + q(t)u \geqslant 0, \qquad u(0) = c_1, \quad u'(0) = c_2, \tag{10.9.2}$$

implies that $u \geqslant v$, *where* v *is the solution of*

$$v'' + p(t)\,v' + q(t)v = 0, \qquad v(0) = c_1\,, \quad v'(0) = c_2\,. \tag{10.9.3}$$

The proof is immediate. Let u_2 be another solution of (10.9.1) which is linearly independent of u_1. From (10.9.2) and (10.9.3), we have

$$(u - v)'' + p(t)(u - v)' + q(t)(u - v) = r(t) \geqslant 0. \tag{10.9.4}$$

Using the identity of (10.8.6), this may be written

$$\frac{d}{dt}\left(\frac{u_1^2}{W(u_1\,,\, u_2)}\frac{d}{dt}\left(\frac{u - v}{u_1}\right)\right) = \frac{u_1 r(t)}{W(u_1\,,\, u_2)} \geqslant 0. \tag{10.9.5}$$

Integrating between 0 and t, we have

$$\frac{u_1^2}{W(u_1\,,\, u_2)}\frac{d}{dt}\left(\frac{u - v}{u_1}\right) = \int_0^t \frac{u_1 r(s)\, ds}{W(u_1\,,\, u_2)} \geqslant 0. \tag{10.9.6}$$

The nonnegativity of the right-hand side follows from the positivity of u_1, the nonnegativity of r, and the positivity of $W(u_1\,,\, u_2)$. Recall the Abel identity derived in Sec. 1.4, Volume I.

From (10.9.6), we have, integrating once again,

$$\frac{u - v}{u_1} \geqslant \int_0^t \frac{W(u_1\,,\, u_2)}{u_1^2}\left[\int_0^{t_1} \frac{u_1 r(s)\, ds}{W(u_1\,,\, u_2)}\right] dt_1\,. \tag{10.9.7}$$

Thus, given a measure of the positivity of $L(u - v)$, the function $r(t)$, we can provide a measure of the excess of u over v.

10.10. A Further Condition

In place of imposing the condition that there exist a solution u_1 which is positive in $[0, t_0]$, we can ask that the Riccati equation

$$r' + r^2 + p(t)r + q(t) = 0 \tag{10.10.1}$$

possess a solution in the interval $[0, t_0]$. The conditions are equivalent, since the change of variable $w'/w = r$ transforms (10.10.1) into the second-order linear differential equation we have been treating. Furthermore, for r to exist over $[0, t_0]$, it is necessary that w be nonzero in this interval.

10.11. Two-point Boundary Conditions

Prior to a consideration of corresponding results for the nth-order linear differential equation, let us consider the differential inequality

$$u'' + p(t)\,u' + q(t)u \leqslant 0, \qquad 0 \leqslant t \leqslant b, \tag{10.11.1}$$

where u is now constrained by a two-point boundary condition $u(0) = u(b) = 0$. As before, it is sufficient to suppose that $p(t) = 0$, and eliminate the term $p(t)u'$.

The inequality is then equivalent to the differential equation

$$u'' + q(t)u + f(t) = 0, \qquad u(0) = u(b) = 0, \tag{10.11.2}$$

where $f(t) \geqslant 0$.

Since the solution to this equation is

$$u = \int_0^b k(t, t_1) f(t_1)\, dt_1\,, \tag{10.11.3}$$

we see that everything depends on the nature of the Green's function $k(t, t_1)$.

We will present two approaches to the establishment of the positivity of k. The first is based upon an associated variational problem, the second upon an associated partial differential equation.

10.12. Variational Approach

We begin with the observation, previously made, that the equation of (10.11.2) is the Euler equation associated with the problem of minimizing the quadratic functional

$$J(u) = \int_0^b [u'^2 - q(t)\,u^2 - 2f(t)u]\, dt, \tag{10.12.1}$$

subject to the condition that $u(0) = u(b) = 0$. Let us impose the condition that the quadratic functional

$$J_1(u) = \int_0^b [u'^2 - q(t)\,u^2]\, dt \tag{10.12.2}$$

is positive definite for all admissible u; i.e., $u' \in L^2(0, b)$, $u(0) = u(b) = 0$.

Then the minimum of $J(u)$ exists and is furnished by the unique solution of (10.11.2), as we have already demonstrated in Chapter 8,

Volume I. Let us now show that this function must be nonnegative in $[0, b]$ if $f(t) \geqslant 0$ in this interval. Consider Fig. 10.2. Suppose that there exists an interval $[a_1, a_2]$, as indicated, within which $u \leqslant 0$. Consider then the new function u_1 defined as follows:

$$\begin{aligned} u_1 &= u, \qquad && t \text{ outside of } [a_1, a_2], \\ &= -u, && a_1 \leqslant t \leqslant a_2 . \end{aligned} \tag{10.12.3}$$

This is an admissible function since $u_1' \in L^2(0, b)$. It is clear, however, that $J(u_1) < J(u)$, a contradiction to the minimizing property of u.

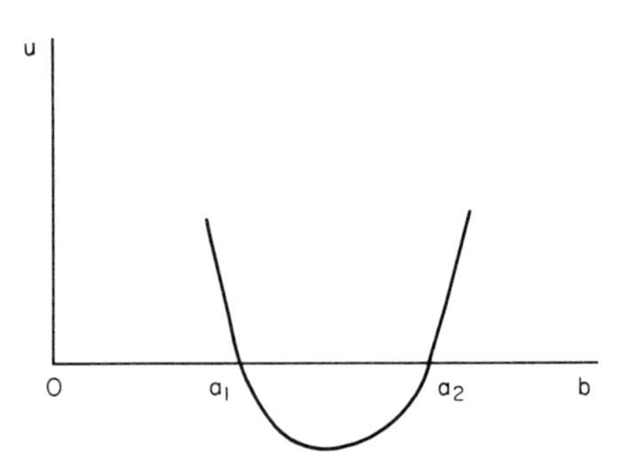

Figure 10.2

If $u \geqslant 0$ for arbitrary $f \geqslant 0$, it follows that $k(t, t_1) \geqslant 0$ for $0 \leqslant t_1$, $t \leqslant b$. Hence, we have established

Theorem. *If the quadratic form*

$$J_1(u) = \int_0^b (u'^2 - q(t)\, u^2)\, dt \tag{10.12.4}$$

is positive definite for all $u' \in L^2(0, b)$ such that $u(0) = u(b) = 0$, then the Green's function associated with the equation

$$u'' + q(t)u = -f(t), \qquad u(0) = u(b) = 0, \tag{10.12.5}$$

is nonnegative.

Exercises

1. Extend the foregoing argument to show that $u(t) = \int_0^b k(t, t_1) f(t_1)\, dt_1$ is a variation-diminishing transformation in the sense that $u(t)$

cannot have more changes of sign than the continuous function $f(t)$. See

R. Bellman, "On Variation-diminishing Properties of Green's Functions," *Boll. D'unione Math.*, Vol. 16, 1961, pp. 164–166.

2. Extend the foregoing argument to the equation $u'' + q_1(t)u - q_2(t)u^3 = f(t)$, $u(0) = u(b) = 0$, under appropriate assumptions concerning the functional $\int_0^b [u'^2 - q_1u^2 + q_2u^4/2]\, dt$.

10.13. A Related Parabolic Partial Differential Equation

Let us now consider the behavior of the solution of

$$u'' + q(t)u + f(t) = 0, \qquad u(0) = u(b) = 0, \tag{10.13.1}$$

by considering it to be the limiting form of the partial differential equation

$$\frac{\partial u}{\partial s} = \frac{\partial^2 u}{\partial t^2} + q(t)u + f(t), \qquad u(0, s) = u(b, s) = 0, \quad u(t, 0) = g(t), \tag{10.13.2}$$

as $s \to \infty$.

Suppose that we can show that $u(t, s) \geqslant 0$ for $0 \leqslant t \leqslant b$, $s \geqslant 0$, under the assumption that the initial function $g(t)$ is nonnegative, and suppose further that we can show that $\lim_{s\to\infty} u(t, s) = u(t)$, the solution of (10.13.1). If so, it follows that $u(t) \geqslant 0$.

Let us sketch the proof without filling in all of the details. We are interested primarily in indicating some of the interconnections between ordinary and partial differential equations, and incidentally in pointing out some of the virtues of imbedding techniques.

10.14. Nonnegativity of $u(t, s)$

Consider, to begin with, the heat equation

$$u_s = u_{tt} + f(t). \tag{10.14.1}$$

Let us examine a random walk equation for which (10.14.1) is the limiting form, namely

$$w\left(t, s + \frac{\Delta^2}{2}\right) = \frac{w(t + \Delta, s) + w(t - \Delta, s)}{2} + \frac{f(t)\,\Delta^2}{2}, \tag{10.14.2}$$

where t assumes the discrete values $0, \Delta, 2\Delta, \ldots$, and s the discrete values $0, \Delta^2/2, \ldots$. It is clear from this recurrence relation that $w(t, s) \geqslant 0$ if $w(t, 0) \geqslant 0$, $f(t) \geqslant 0$, and hence that the limiting function obtained as $\Delta \to 0$ is nonnegative, provided, of course, that it exists. We shall refer the reader to other sources for a discussion of when this convergence occurs. We have previously indicated that this is a stability question concerning the solution of (10.14.2) once we have established the existence of a solution of (10.14.1) with sufficient smoothness.

To handle the original equation,

$$u_s = u_{tt} + q(t)u + f(t), \tag{10.14.3}$$

we first suppose that $q(t) \geqslant 0$. The difference equation technique used above then shows that $u \geqslant 0$ for $s \geqslant 0$, $0 \leqslant t \leqslant b$.

The restriction that $q \geqslant 0$ is unessential since the change of variable

$$u = e^{-kt}v \tag{10.14.4}$$

converts (10.14.3) into

$$v_s = v_{tt} + (q(t) + k)v + f(t)\, e^{kt}. \tag{10.14.5}$$

Hence, it is sufficient to assume that

$$q(t) + k \geqslant 0 \tag{10.14.6}$$

for some constant k, and even a weaker condition that

$$q(t) + k(t) \geqslant 0 \tag{10.14.7}$$

for some function $k(t)$ such that $\int_0^t k(t_1)\, dt_1$ exists for $t \geqslant 0$.

Assuming the legitimacy of the use of difference equations, we have demonstrated

Theorem. *If*

$$\begin{aligned}
&\text{(a)}\quad u_t - u_{xx} - q(x)u \geqslant 0, \qquad 0 < x < b, \quad t > 0,\\
&\text{(b)}\quad u(x, 0) \geqslant 0, \qquad 0 \leqslant x \leqslant b,\\
&\text{(c)}\quad u(0, t) = u(b, t) = 0, \qquad t \geqslant 0,\\
&\text{(d)}\quad q(x) \geqslant -k > -\infty, \qquad 0 \leqslant x \leqslant b, \quad 0 \leqslant t \leqslant T, \quad \textit{for any} \quad T > 0,
\end{aligned} \tag{10.14.8}$$

then $u(x, t) \geqslant 0$ *for* $0 \leqslant x \leqslant b$, $t \geqslant 0$.

10.15. Limiting Behavior

It remains to investigate conditions under which the solution of the partial differential equation

$$\begin{aligned} u_t &= u_{xx} + q(x)u + f(x), \\ u(0, t) &= u(1, t) = 0, \qquad t > 0, \\ u(x, 0) &= h(x), \qquad 0 \leqslant x \leqslant 1, \end{aligned} \tag{10.15.1}$$

converges to the solution of the ordinary differential equation

$$v'' + q(x)v + f(x) = 0, \qquad v(0) = v(1) = 0, \tag{10.15.2}$$

as $t \to \infty$. We have reverted to the usual space and time variables for the heat equation.

If we set $u = v + w$, we see that w satisfies

$$\begin{aligned} w_t &= w_{xx} + q(x)w, \\ w(0, t) &= w(1, t) = 0, \qquad t > 0, \\ w(x, 0) &= h(x) - v(x). \end{aligned} \tag{10.15.3}$$

Hence, we want to know when $w \to 0$ as $t \to \infty$, regardless of the nature of $h(x) - v(x)$.

The first clue to an answer is provided by a Sturm–Liouville expansion,

$$w = \sum_{k=1}^{\infty} e^{\lambda_k t} v_k(x), \tag{10.15.4}$$

where the λ_k and v_k are determined by the equation

$$\lambda_k v_k - v_k'' - q(x)\, v_k = 0, \qquad v_k(0) = v_k(1) = 0, \tag{10.15.5}$$

with the v_k normalized by the condition $\int_0^1 {v_k}^2\, dx = 1$.

If all the characteristic values are negative, $\lambda_k < 0$, we see that $w \to 0$ as $t \to 0$. Integrating (10.15.5), after multiplication by v_k, we see that

$$\lambda_k \int_0^1 {v_k}^2\, dx = \int_0^1 v_k(v_k'' + q(x)\, v_k)\, dx = \int_0^1 (q{v_k}^2 - v_k'^2)\, dx < 0 \tag{10.15.6}$$

if the quadratic functional $\int_0^1 (u'^2 - qu^2)\, dx$ is positive definite. Once we see the nature of the desired answer, it turns out we can establish the result without the necessity of using the theory of Sturm–Liouville expansions.

10.16. Limiting Behavior: Energy Inequalities

To obtain the desired result in a simpler fashion, we employ an important method in the theory of partial differential equations, the so-called "energy integral." This is related to the method used by Lyapunov to study questions of stability as we indicate below.

Let us illustrate the technique starting with the simpler equation

$$\begin{aligned} u_t &= u_{xx}, \\ u(0, t) &= u(1, t) = 0, \qquad t > 0, \\ u(x, 0) &= h(x). \end{aligned} \tag{10.16.1}$$

Considering the functional $J(u) = \int_0^1 u^2\, dx$, we have

$$\begin{aligned} \frac{d}{dt}\left(\int_0^1 u^2\, dx\right) &= 2\int_0^1 uu_t\, dx = 2\int_0^1 uu_{xx}\, dx \\ &= 2uu_x\Big]_0^1 - 2\int_0^1 u_x^2\, dx = -2\int_0^1 u_x^2\, dx. \end{aligned} \tag{10.16.2}$$

Hence, we see that $J(u)$ is monotone decreasing as a function of t. However, we can go much further. By virtue of Wirtinger's inequality (discussed in Sec. 8.11, Volume I) we have

$$\int_0^1 u_x^2\, dx \geqslant \pi^2 \int_0^1 u^2\, dx, \tag{10.16.3}$$

for any function u vanishing at $x = 0$ and 1. Hence, we see that

$$\frac{d}{dt}\left(\int_0^1 u^2\, dx\right) \leqslant -2\pi^2 \int_0^1 u^2\, dx, \tag{10.16.4}$$

whence for $t \geqslant 0$

$$\int_0^1 u^2\, dx \leqslant \exp(-2\pi^2 t) \int_0^1 u(x, 0)^2\, dx. \tag{10.16.5}$$

If the basic equation is

$$u_t = u_{xx} + q(x)u, \tag{10.16.6}$$

we obtain via the procedure in (10.16.2) the equation

$$\begin{aligned} \frac{d}{dt}\left(\int_0^1 u^2\, dx\right) &= 2\int_0^1 uu_t\, dx = 2\int_0^1 u(u_{xx} + q(x)u)\, dx \\ &= -2\int_0^1 (u_x^2 - q(x)\, u^2)\, dx. \end{aligned} \tag{10.16.7}$$

If $\int_0^1 (u_x^2 - q(x)u^2)\,dx$ is positive definite for functions vanishing at $x = 0$ and 1, we see that

$$\int_0^1 (u_x^2 - q(x)\,u^2)\,dx \geqslant \lambda_1 \int_0^1 u^2\,dx \tag{10.16.8}$$

for some $\lambda_1 > 0$. Hence

$$\frac{d}{dt}\left(\int_0^1 u^2\,dx\right) \leqslant -2\lambda_1 \int_0^1 u^2\,dx, \tag{10.16.9}$$

whence

$$\int_0^1 u^2\,dx \leqslant e^{-2\lambda_1 t} \int_0^1 u(x,0)^2\,dx. \tag{10.16.10}$$

This makes it plausible that $u \to 0$ as $t \to \infty$, but does not establish it rigorously. To do this, we turn to the original equation, (10.16.6), and consider it for $t \geqslant t_0 > 0$. Then $w = u_t$ satisfies the equation

$$\begin{aligned} w_t &= w_{xx} + q(x)w, \\ w(0,t) &= w(1,t) = 0, \\ w(x,t_0) &= h_1(x), \qquad h_1(x) = u_t(x,t_0). \end{aligned} \tag{10.16.11}$$

Applying the same reasoning as above to this equation, we have

$$\int_0^1 u_t^2\,dx \leqslant e^{-2\lambda_1 t} \int_0^1 u_t(x,t_0)^2\,dx \tag{10.16.12}$$

for $t \geqslant t_0$. Hence, $\int_0^1 u_t^2\,dx \to 0$ as $t \to \infty$.

Returning to (10.16.6), we write

$$u = -\int_0^1 k(x,x_1)\,u_t\,dx_1, \tag{10.16.13}$$

where $k(x,x_1)$ in the Green's function associated with $u'' + q(x)u$ and the boundary conditions $u(0) = u(1) = 0$, whence

$$|u| \leqslant \left(\int_0^1 k^2(x,x_1)\,dx_1\right)^{1/2} \left(\int_0^1 u_t^2\,dx_1\right)^{1/2}. \tag{10.16.14}$$

Hence $u \to 0$ as $t \to \infty$.

Exercise

1. Show in the same way that $u_t, u_{tt}, \ldots,$ all approach zero as $t \to \infty$, and at the same rate.

10.17. Monotonicity of Maximum

The same simple techniques establish the fact that the functional

$$I_n = \int_0^1 u^{2n}\, dx \tag{10.17.1}$$

is a decreasing function of t for $t \geqslant 0$ if u satisfies the equation of (10.16.1).

Since

$$\lim_{n\to\infty} I_n^{1/2n} = \max_{0\leqslant x\leqslant 1} |\, u \,|, \tag{10.17.2}$$

we see that the maximum value of u is monotone decreasing.

If

$$k(x)\, u_t = u_{xx}\,, \tag{10.17.3}$$

where $0 < k(x) < k_0$, we obtain the same result by considering the modified functional

$$I_n = \int_0^1 u^{2n} k(x)\, dx. \tag{10.17.4}$$

Exercise

1. Apply the foregoing method to study the equation

$$u_t = g(u)\, u_x + u_{xx}$$

under the boundary conditions $u(0, t) = u(1, t) = 0, \quad t > 0$, $u(x, 0) = h(x)$.

10.18. Lyapunov Functions

Use of the foregoing ideas provides us with an approach to the study of the stability of solutions of nonlinear differential equations which is both more versatile and more flexible than that presented in Chapter 2, Volume I. In return, it is less precise. In many cases, however, it is the only method we possess.

The linear approximation

$$\frac{dy}{dt} = Ay, \qquad y(0) = c, \tag{10.18.1}$$

is used to study the stability of the solution $x = 0$ of

$$\frac{dx}{dt} = Ax + g(x), \tag{10.18.2}$$

by means of an examination of the explicit solution of (10.18.1) and the associated inhomogeneous equation. It turns out that it is far more convenient in many cases to pursue a different approach, the celebrated second method of Lyapunov. It is a natural approach to a physicist or engineer since it depends upon the systematic use of energy integrals.

Consider the full nonlinear equation

$$\frac{dx}{dt} = Ax + g(x), \qquad x(0) = c, \tag{10.18.3}$$

and suppose that we can find a continuous scalar function $V(x)$ of the vector x such that $V(x) > 0$ for $x \neq 0$, $V(0) = 0$, and such that in addition $V(x(t))$ approaches 0 as $t \to \infty$, where $x(t)$ is the solution of (10.18.3). It follows that x approaches 0 as $t \to \infty$, the desired stability.

To obtain a function of this type often requires a good deal of ingenuity and experimentation. One technique, following Lyapunov himself, is to look for a function V satisfying the scalar differential inequality

$$\frac{dV}{dt} \leqslant -kV, \tag{10.18.4}$$

$k > 0$, whenever x satisfies (10.18.3). It is clear from this that $V \leqslant V(0)e^{-kt}$ and thus that $V \to 0$ as $t \to \infty$.

In order to illustrate how this function V is obtained, consider the trial function

$$V = (x, Bx), \tag{10.18.5}$$

where B is a positive definite matrix to be chosen conveniently. We have, using (10.18.3),

$$\begin{aligned} \frac{dV}{dt} &= (Ax + g(x), Bx) + (x, B(Ax + g(x))) \\ &= ((BA + A'B)x, x) + 2(g(x), Bx). \end{aligned} \tag{10.18.6}$$

Here A' denotes the transpose of A.

Let us now attempt to choose B so that

$$BA + A'B = -I. \tag{10.18.7}$$

As we have seen in Sec. 3.19, Volume I, the existence of a unique positive definite matrix B satisfying this equation is a necessary and sufficient condition for all the characteristic roots of A to have negative real parts, i.e., for A to be a stability matrix. In that case, an explicit representation of B is given very simply by the integral

$$B = \int_0^\infty e^{At} e^{A't}\, dt. \tag{10.18.8}$$

With this choice of B, the equation in (10.18.6) becomes

$$\frac{dV}{dt} = -V + 2(g(x), Bx). \tag{10.18.9}$$

If we impose the stability condition $\| g(x)\| \leqslant \epsilon \| x \|$ for $\| x \|$ small, we see that

$$|(g(x), Bx)| \leqslant \epsilon_2(x, x) \leqslant \epsilon_1(x, Bx) = \epsilon_1 V \tag{10.18.10}$$

for x small, where ϵ_1 is a small quantity if ϵ_2 is. Thus, (10.18.9) reads

$$\frac{dV}{dt} \leqslant (-1 + \epsilon_1)V, \qquad V(0) = (c, Bc), \tag{10.18.11}$$

for $\| c \|$ small and some initial t-interval $[0, t_0]$. Hence, for $0 \leqslant t \leqslant t_0$,

$$V \leqslant V(0)\, e^{(-1+\epsilon_1)t} \leqslant V(0). \tag{10.18.12}$$

It follows that $V(x)$ is monotonically decreasing as a function of t if $\epsilon_1 < 1$. Thus, if $V(0) = (c, Bc)$ is small, we know that (x, Bx) is small for $t \geqslant 0$. Since B is positive definite, this implies that (x, x) remains small. Consequently, if c is chosen so that (10.18.10) holds at $t = 0$ with $\epsilon_1 < 1$ the inequality holds for all $t \geqslant 0$.

This establishes stability. Observe that the qualitative property has been studied with only the most rudimentary quantitative considerations. It is this which makes the Lyapunov technique so versatile. In particular, it can be used for design purposes where the choice of components in a physical system must be made to ensure stability.

Exercise

1. Consider the equations $u' = -a_{11}u + a_{12}v + b_1 uv$, $u(0) = c_1$, $v' = a_{21}u - a_{22}v + b_2 uv$, $v(0) = c_2$, where $a_{ij} \geqslant 0$, $b_1, b_2 \geqslant 0$, $c_1, c_2 \geqslant 0$. If $|c_1|$, $|c_2|$ are sufficiently small and the associated characteristic roots have negative real parts, the solutions tend to zero as $t \to \infty$.

Suppose that it is known that $0 \leqslant u \leqslant k_1$ and $0 \leqslant v \leqslant k_2$. Show that by considering the associated linear equation $u' = -a_{11}u + a_{12}v + b_1k_2u$, $v' = a_{21}u - a_{22}v + b_2k_1v$, $u(0) = c_1$, $v(0) = c_2$, the conditions on smallness of $|c_1|$ and $|c_2|$ can be removed if

$$\begin{pmatrix} -a_{11} + b_1k_2 & a_{12} \\ a_{21} & -a_{22} + b_2b_{k_1} \end{pmatrix}$$

is a stability matrix. The results of Sec. 10.22 will be helpful. See

R. Bellman, "Vector Lyapunov Functions," *J. SIAM Control*, Vol. 1, 1962, pp. 32–34.

For further material on vector Lyapunov functions, see

V. Lakshmikantham and S. Leela, *Differential and Integral Inequalities*, Vol. I, Academic Press, New York, 1969.

10.19. Factorization of the *n*th-order Linear Operator

Let us now establish the n-dimensional version of the identity used in Sec. 10.8 to obtain the factorization of the second order linear differential operator. Introduce the determinant functions

$$U_r = \begin{vmatrix} u & u_1 & \cdots & u_r \\ u' & u_1{}' & \cdots & u_r{}' \\ \vdots & & & \\ u^{(r)} & u_1^{(r)} & \cdots & u_r^{(r)} \end{vmatrix},$$

$$W_r = \begin{vmatrix} u_1 & u_r & \cdots & u_r \\ \vdots & & & \\ u_1^{(r-1)} & u_2^{(r-1)} & \cdots & u_r^{(r-1)} \end{vmatrix}, \tag{10.19.1}$$

for $r = 1, 2,\ldots, n$, under the assumption that the u_i are n linearly independent solutions of the linear differential equation

$$L(u) = u^{(n)} + p_1(t)\, u^{(n-1)} + \cdots + p_n(t)u = 0. \tag{10.19.2}$$

These are Wronskians. We now suppose that the n linearly independent solutions are such that

$$u_1 > 0, \quad W_2(u_1, u_2) > 0,\ldots, W_n(u_1, u_2,\ldots, u_n) > 0 \tag{10.19.3}$$

is an interval $[0, t_0]$.

This is called "property W."

Repeating a previous approach, we see that we can write the equation in (10.19.2) in the form

$$L(u) = U_n/W_n = 0. \tag{10.19.4}$$

Let us now establish the identity

$$U_r W_{r-1} = U_{r-1} W_r' - U_{r-1}' W_r \tag{10.19.5}$$

for $r = 1, 2, \ldots$. Here $W_0 = 1$, $U_0 = u$. This can be proved by determinantal manipulation, but it is far simpler to establish it by using the fact that both sides of (10.19.5) are linear differential expressions which vanish for $u = u_1, \ldots, u_r$, and that the coefficients of $u^{(r)}$ on both sides are equal. We are leaning on the result that a linear differential equation is uniquely determined by its solutions once the coefficient of the highest derivative is fixed.

It follows that

$$\begin{aligned} L(u) &= \frac{U_n}{W_n} = \left(\frac{W_n}{W_{n-1}}\right)\left(\frac{W_{n-1}U_n}{W_n^2}\right) \\ &= -\frac{W_n}{W_{n-1}} \frac{d}{dt}\left(\frac{U_{n-1}}{W_n}\right) \\ &= -\frac{W_n}{W_{n-1}} \frac{d}{dt}\left(\frac{W_{n-1}}{W_n}\left(\frac{U_{n-1}}{W_{n-1}}\right)\right). \end{aligned} \tag{10.19.6}$$

Iterating this relation, we have the desired decomposition.

$$L(u) = (-1)^n \frac{W_n}{W_{n-1}} \frac{d}{dt}\left(\frac{W_{n-1}^2}{W_n W_{n-2}}\left(\frac{d}{dt} \cdots \frac{d}{dt}\left(\frac{W_1^2}{W_0 W_2}\left(\frac{d}{dt}\left(\frac{W_0 u}{W_1}\right) \cdots\right) = 0 \tag{10.19.7}$$

for $0 \leqslant t \leqslant t_0$ under the assumption that property W is satisfied in this interval.

Exercises

1. Show that if $u_1, \ldots, u_n$ constitute a principal system, we can write

$$u_1 = v_1, \quad u_2 = v_1 \int v_2\, dx, \ldots, u_n = v_1 \int v_2 \int \cdots \int v_n (dt)^{n-1},$$

and (10.19.7) becomes

$$\frac{d}{dt}\left(\frac{1}{v_n} \frac{d}{dt}\left(\frac{1}{v_{n-1}} \frac{d}{dt} \cdots \frac{d}{v_2 dt}\left(\frac{u}{v_1}\right) \cdots\right) = 0.$$

2. Use Rolle's theorem and the foregoing factorization of the linear operator to prove that if $f(t)$ is differentiable n times and vanishes at n points in the interval (a, b), then there exists an intermediate point s such that $Lf(s) = f^{(n)}(s) + p_1(s)f^{(n-1)}(s) + \cdots + p_n(s)f(s) = 0$, assuming that property W is satisfied in (a, b). This and the following results are taken from the paper

   G. Polya, "On the Mean-Value Theorem Corresponding to a Given Linear Homogeneous Differential Equation," *Trans. Amer. Math. Soc.*, Vol. 24, 1922, pp. 312–324.

3. If $f(t)$ assumes at $n + 1$ points of the interval (a, b) the same values as a solution of $L(u) = 0$, then there is an intermediate point s such that $L(f) = 0$.

4. No solution of $L(u) = 1$ can vanish at more than n points of (a, b).

5. Let $a_1 < a_2 < \cdots < a_n$, $t_1 < t_2 < \cdots < t_n$, then

$$\begin{vmatrix} e^{a_1t_1} & e^{a_1t_2} & \cdots & e^{a_1t_n} \\ e^{a_2t_1} & e^{a_2t_2} & \cdots & e^{a_2t_n} \\ \vdots & & & \\ e^{a_nt_1} & e^{a_nt_2} & \cdots & e^{a_nt_n} \end{vmatrix} \neq 0.$$

6. There exists one and only one solution of $L(u) = 1$ assuming n given values at n given points of (a, b).

7. Let $u(t)$ be a solution of $L(u) = 0$ assuming the same values as $f(t)$ at n given points of (a, b); let $v(t)$ be a solution of $L(u) = 1$ vanishing at these n points. Then there is a point y in (a, b) such that

$$f(t) = u(t) + v(t)\,L(f(y)).$$

   This constitutes an extensive generalization of the Taylor formula with a remainder term. See also

   D. V. Widder, "A Generalization of Taylor's Series," *Trans. Amer. Math. Soc.*, Vol. 30, 1928, pp. 126–154.

8. Show that we can write

$$f(t) = f(t_1)\,e^{t_1-t} + (1 - e^{t_1-t})(f(s) + f'(s)),$$

   for some s in (t_1, t). Hence show that

$$\lim_{t\to\infty} (f(t) + f'(t)) = c \qquad \text{implies} \quad \lim_{t\to\infty} f(t) = c.$$

   (For an alternative approach, see Sec. 1.3, Exercise 7, in Volume I.)

9. Show that we may write

$$u(t) = \frac{u(t_1)\sin(t_2 - t) + u(t_2)\sin(t - t_1)}{\sin(t_2 - t_1)} + \left(1 - \frac{\cos(t - \frac{1}{2}(t_1 + t_2))}{\cos^2(\frac{1}{2}(t_2 - t_1))}\right)(u''(y) + u(y))$$

for y in the smallest interval containing t_1, t_2, and t, where $t_1 < t_2 < t_1 + \pi$. Why does π enter?

10. Show that the distance between any two zeros of a solution of $u'' + \varphi(t)u = 0$ where $\varphi(t) > 1$ is less than π. This is a special case of the classical result of Sturm. For general results, see

M. Bocher, *Méthodes de Sturm*, Paris, 1917.

11. Any solution of the foregoing equation has the property that $u(t) > 0$ implies $u''(t) + u(t) < 0$.

12. Can analogous results, decomposition, etc., be obtained for linear difference equations of the form

$$u_n + a_1(n)\, u_{n-1} + \cdots + a_r(n)\, u_{n-r} = 0\,?$$

10.20. A Result for the *n*th-order Linear Differential Equation

We can now state

Theorem. *If $L(u) = 0$ possesses n solutions $u_1, u_2, \ldots, u_n$ satisfying property W in $[0, t_0]$, then*

$$L(u) \leqslant 0, \quad u(0) = c_1, \ldots, u^{(n-1)}(0) = c_n, \tag{10.20.1}$$

in $[0, t_0]$ implies that $u \leqslant v$, where v satisfies the equation

$$L(v) = 0, \quad v(0) = c_1, \ldots, v^{(n-1)}(0) = c_n. \tag{10.20.2}$$

A proof follows immediately from the foregoing decomposition.

10.21. An Example

The equation

$$L(u) = u^{(n)} + p_1 u^{(n-1)} + \cdots + p_n u = 0, \tag{10.21.1}$$

where the p_i are now constants, satisfies property W in every finite interval if the roots of the characteristic polynomial

$$r^n + p_1 r^{(n-1)} + \cdots + p_n = 0 \tag{10.21.2}$$

are all real and distinct. Let these roots be denoted by $\lambda_1, \lambda_2, ..., \lambda_n$. Let $u_k(t) = \exp(\lambda_k t)$. Then it is easy to see that

$$W(u_1, u_2, ..., u_k) = \exp((\lambda_1 + \lambda_2 + \cdots + \lambda_k)t) \begin{vmatrix} 1 & \lambda_1 & \cdots & \lambda_1^{k-1} \\ 1 & \lambda_2 & \cdots & \lambda_2^{k-1} \\ \vdots & & & \\ 1 & \lambda_k & \cdots & \lambda_k^{k-1} \end{vmatrix}.$$

The determinant, the Vandermonde determinant, is nonzero, if the λ_i are all distinct.

Exercise

1. Establish the result that $u \leqslant v$ by use of the factorization

$$L_n(u) = \prod_{i=1}^{n} \left(\frac{d}{dt} - \lambda_i\right) u.$$

10.22. Linear Systems

It appears quite difficult to obtain any extensive results concerning systems of linear differential inequalities of the form

$$\frac{dx}{dt} \leqslant Ax + f(t), \qquad x(0) = c. \tag{10.22.1}$$

We have already seen, in Chapter 3, Volume I, that a necessary and sufficient condition for all elements of e^{At} to be nonnegative for $t \geqslant 0$ is that $a_{ij} \geqslant 0$ for $i \neq j$. Hence we can conclude that if y is the solution of

$$\frac{dy}{dt} = Ay + f(t), \qquad y(0) = c, \tag{10.22.2}$$

and A satisfies the foregoing condition, then $x \leqslant y$.

Exercise

1. Show that we can allow variable $A(t)$ with $a_{ij}(t) \geqslant 0$ for $i \neq j$.

10.23. Partial Differential Equation—I

Let us now consider the question of determining when

$$u_{xx} + u_{yy} + q(x, y)u \leqslant 0 \tag{10.23.1}$$

for $(x, y) \in R$ and $u = f$ on B, the boundary of R, implies that

$$u \geqslant v, \tag{10.23.2}$$

where v is a solution of the corresponding equality.

Two of the methods we gave in the foregoing sections generalize readily, that based upon the variation of

$$J(u) = \int_R (u_x^2 + u_y^2 - q(x, y)\, u^2 + 2f(x, y)u)\, dx\, dy, \tag{10.23.3}$$

and that based upon the use of the associated parabolic partial differential equation

$$w_t = w_{xx} + w_{yy} + q(x, y)w. \tag{10.23.4}$$

10.24. Partial Differential Equation—II

We shall also encounter the equation

$$u_t = g(u, x, t) + h(u, x, t)\, u_x, \qquad u(x, 0) = 0, \tag{10.24.1}$$

in connection with the theory of dynamic programming and invariant imbedding. The unconventional difference equation

$$u(x, t + \Delta) = g(u, x, t)\Delta + u(x + h(u, x, t)\Delta, t), \tag{10.24.2}$$

establishes the required positivity, provided we can show the required limiting behavior as $\Delta \to 0$, as does also the usual solution based upon characteristics.

Miscellaneous Exercises

1. Let $\{x_k\}$, $\{f_k\}$, and $\{z_k\}$, $k = 0, 1, \ldots, m$, be real-valued sequences and let $z_k \geqslant 0$. If, for $k = 0, 1, \ldots, m$, $x_k \leqslant f_k + \sum_{0 \leqslant i < k} f_i x_i$, then

$$x_k \leqslant f_k + \sum_{0 \leqslant i \leqslant k} \Big[\prod_{i<j<k} (1 + z_j) f_i f_i \Big].$$

2. Let x, f, g, and z be real-valued functions defined on an interval $[a, b]$ and either continuous or of bounded variation. Let g and z be nonnegative and let h be a nondecreasing continuous function defined in $[a, b]$. If for all t in $[a, b]$,

$$x(t) \leqslant f(t) + g(t) \int_a^t z(t_1)\, x(t_1)\, dh(t_1),$$

then

$$x(t) \leqslant f(t) + g(t) \int_a^t f(t_1)\, z(t_1) \exp\left(\int_{t_1}^t g(s)\, z(s)\, dh(s)\, dh(t_1)\right)$$

for $a \leqslant t < b$.

G. S. Jones, "Fundamental Inequalities for Discrete and Discontinuous Functional Equations," *J. SIAM*, Vol. 12, 1964, pp. 43–57.

3. From $u_{xx} - u_t - gu - f = 0$, $u(x, 0) = 0$, $u(0, t) = u(\pi, t) = 0$ conclude that

$$\frac{d}{dt}\left(\int_0^\pi u^2\, dx\right) = -\int_0^\pi (u_x{}^2 + gu^2)\, dx - \int_0^\pi fu\, dx \leqslant -\int_0^\pi fu\, dx,$$

and hence that

$$\frac{d}{dt}\left(\int_0^\pi u^2\, dx\right) \leqslant \left(\int_0^\pi f^2\, dx\right)^{1/2} \left(\int_0^\pi u^2\, dx\right)^{1/2}.$$

4. Show that if $v(t)$ is a nonnegative function satisfying the inequality $v'(t) \leqslant av(t)^{1/2}$ for $t \geqslant 0$ for some $a \geqslant 0$, then $v(t)^{1/2} \leqslant v(0)^{1/2} + at/2$.

5. If $v(t)$ is a nonnegative function satisfying the inequality $v'(t) \leqslant a(t)v(t)^{1/2}$ for $t \geqslant 0$, where $a(t)$ is nonnegative and nondecreasing, then $v(t)^{1/2} \leqslant v(0)^{1/2} + ta(t)/2$.

6. Combining the three foregoing results, show that

$$\int_0^\pi u^2(x, t)\, dt \leqslant t^2 \left[\max_{0 \leqslant t} \left(\int_0^\pi f^2(x, t)\, dx\right)^{1/2}\right]^2.$$

7. The function u_t satisfies the equation

$$w_{xx} - w_t - gw - g_t u - f_t = 0,$$

obtained by differentiation of the equation for u. Hence, obtain a bound for $\int_0^\pi u_t{}^2\, dx$.

8. Using the relation $\int_0^\pi (-u_x{}^2 - uu_t - gu^2 - fu)\,dx = 0$, show that $\int_0^\pi u_x{}^2\,dx$ can be majorized by terms involving $\int_0^\pi u^2\,dx$, $\int_0^\pi u_t{}^2\,dx$.

9. If $f(t) \geqslant 0$, $0 \leqslant t \leqslant a$, and

$$kf(t)^m \geqslant (a-t)^{-p} \int_t^a (a-s)^q f(s)^n\,ds > 0$$

for all t in $[0, a]$, with $k, m, n, p, q > 0$, then $f'(0) \geqslant a/c_1$. See

R. Bellman, "On a Differential Inequality of Cesari and Turner," *Rend. Circ. Math. Palermo*, Serie 11-Tomo VII, 1958, pp. 1–3.

10. If f is continuous and

$$f(\lambda x + (1-\lambda)y) \leqslant \epsilon + \lambda f(x) + 1 - \lambda) f(y)$$

for $0 \leqslant \lambda \leqslant 1$, and all x, y in a convex domain, then there exists a convex function $g(x)$ such that

$$g(x) \leqslant f(x) \leqslant g(x) + k_n \epsilon$$

for a constant k_n, where n is the dimension of x. See

D. H. Hyers and S. M. Ulam, "Approximately Convex Functions," *Proc. Bull. Amer. Math. Soc.*, Vol. 3, 1952, pp. 821–828.

J. W. Green, "Approximately Convex Functions," *Duke Math. J.*, Vol. 19, 1952, pp. 499–504.

11. Let $L(u) = p_0(t)\,u^{(n)} + p_1(t)\,u^{(n-1)} + \cdots + p_n(t)u$. Show that if no nontrivial solution of $L(u) = 0$ can have more than $(n-1)$ zeros in $[a, b]$, counting multiplicities, then $L(u)$ can be written

$$a_0(t)\frac{d}{dt}\left(a_1(t)\frac{d}{dt}\cdots\right)u.$$

Bibliography and Comments

For extensive reference to this area, and, in particular, to "maximum principles" in partial differential equations, see

E. F. Beckenbach and R. Bellman, *Inequalities*, Springer–Verlag, Berlin, 1961,

W. Walter, *Differential-und-Integral-Ungleichungen*, Springer–Verlag, Berlin, 1964,

L. Collatz, "Applications of the Theory of Monotone Operators to Boundary-Value Problems," *Boundary Problems in Differential Equations*, R. E. Langer (ed.), Univ. of Wisconsin Press, Madison, Wisconsin, 1960,

where reference to earlier work will be found.

V. Lakshmikantham and S. Leela, *Differential and Integral Inequalities*, Vol. I and II, Academic Press, New York, 1969.

J. Schroder, "On Linear Differential Inequalities," *J. Math. Anal. Appl.*, Vol. 22, 1968, pp. 188–216.

M. A. Krasnoselskii, *Positive Solutions of Operator Equations* (English translation), Noordhoff, Groningen, 1964.

R. M. Redheffer, "Die Collatzsche Monotonic bei Anfangswert-problemen," *Arch. Rational Mech. Anal.*, Vol. 14, 1963, pp. 196–212.

R. M. Redheffer, "Bemerkungen und Fehlerabschätzung bei Nicht-linearen Partiellen Differentialgleichungen," *Arch. Rational Mech. Anal.*, Vol. 10, 1962, pp. 427–457.

J. Szarski, Differential Inequalities, *Polska Akad. Nauk, Monogr. Mat.*, Tom. 43, Warsaw, 1965.

For applications to ordinary differential equations, see

C. Corduneanu, *Principles of Differential and Integral Equations*, Allyn and Bacon, Rockleigh, New Jersey, 1971.

For an important survey of results for nonlinear equations, see

R. I. Kachurovskii, "Nonlinear Monotone Operators in Banach Spaces," *Russian Math. Surveys*, Vol. 23, 1968, pp. 117–166.

For the work of Caplygin, see

S. A. Caplygin, *New Methods in the Approximate Integration of Differential Equations* (in Russian), Moscow, 1950.

It is occasionally possible to split an operator into two monotone operators. See

O. L. Mangasarian, "Convergent Generalized Monotone Splitting of Matrices," *Math. Comp.*, Vol. 25, 1971, pp. 649–654,

where other references to work of Collatz and Schroder may be found.

§10.3. The fundamental lemma has been discovered independently and used by a number of mathematicians: Let us cite Gronwall, Caligo, Guilano, Reid, Titchmarsh, and Weyl, as well as the author. Its importance in stability theory was first indicated in R. Bellman, "The Stability of Solutions of Linear Differential Equations," *Duke Math. J.*, Vol. 10, 1943, pp. 643–647.

See also

I. Bihari, "A Generalization of a Lemma of Bellman and Its Application to Uniqueness Problems of Differential Equations," *Acta Math. Acad. Sci. Hungar.*, Vol. 7, 1956, pp. 81–94.

C. E. Langenhops, "Bounds on the Norm of a Solution of a General Differential Equation," *Proc. Amer. Math. Soc.*, Vol. 11, 1960, pp. 795–799.

A recent result is

V. N. Laptinskii, "The Gronwall–Bellman Differential Matrix Inequality," *Ukraine Mat. Z.*, Vol. 22, 1970, pp. 690–691.

§10.4. For a detailed study of fourth-order equations, see

J. Schroder, "Randwert Aufgaben vierter Ordnung mit positiver Greenscher Funktion," *Math. Z.*, Vol. 96, 1965, pp. 429–440.

J. Schroder, "Zusammenhangende Mengen Inverspositiver Differential Operatoren Vierter Ordnung," *Math. Zeit.*, Vol. 96, 1967, pp. 89–110.

J. Schroder, "Hinreichende Bedingungen bei Differential-ungleichungen Vierter Ordnung," *Math. Zeit.*, Vol. 92, 1966, pp. 75–94.

J. Schroder, "Operator-ungleichungen und ihre Numerische Anwendung bie Randwertaufgaben," *Numer. Math.*, Vol. 9, 1966, pp. 149–162.

§10.13. For a discussion of some nonlinear problems where this approach is useful, see

D. S. Cohen, "Positive Solutions of a Class of Nonlinear Eigenvalue Problems," *J. Math. Mech.*, August 1967.

D. S. Cohen, "Positive Solutions of a Class of Nonlinear Eigenvalue Problems; Applications to Nonlinear Reactor Dynamics, "*Arch. Rat. Mech. Anal.*, Vol. 26, 1967, pp. 305–315.

§10.14. See, for example,

F. John, "On Integration of Parabolic Equations by Difference Methods—*I*: Linear and Quasilinear Equations for the Infinite Interval," *Comm. Pure Appl. Math.*, Vol. 5, 1952, pp. 155–211.

§10.17. This follows

R. Bellman, "A Property of Summation Kernels," *Duke Math. J.*, Vol. 15, 1948, pp. 1013–1019.

For extensive generalizations, see

S. Bochner, "Quasi-analytic Functions, Laplace Operators, and Positive Kernels," *Ann. Math.*, Vol. 51, 1950, pp. 68–91.

§10.18. For an alternative to the use of Lyapunov functions, see

J. Moser, "On Non-oscillating Networks," *Quart. Appl. Math.*, Vol. 25, 1967, pp. 1–9.

where the Popov technique is used. For the Liapunov method, see

J. P. LaSalle and S. Lefschetz, *Stability by Liapunov's Direct Method*, Academic Press, New York, 1961.

See also,

J. H. George and W. G. Sutton, "Application of Liapunov Theory to Boundary Value Problems," *Proc. Amer. Math. Soc.*, Vol. 25, 1970, pp. 666–671.

J. H. George, V. M. Sehgal, and R. E. Smithson, "Applications of Liapunov's Direct Method to Fixed Point Theorems," *Proc. Amer. Math. Soc* , Vol 28, 1971, pp. 613–620.

For Liapunov's first method, see

N. P. Erugin, "Liapunov's First Method," *Differencial'nye Vravnenija*, Vol. 3, 1967, pp. 531–578.

§10.19. See, for some further results

P. Hartman, "On Logarithmic Derivatives of Solutions of Disconjugate Linear *n*th Order Differential Equations," in *Ordinary Differential Equations*, 1971 NRL-MRC Conf., L. Weiss (ed.), Academic Press, New York, 1972.

A great deal of recent work has been devoted to the study of functions satisfying norm inequalities of the form

$$\left\| \frac{du}{dt} - A(t)\, u(t) \right\|^2 \leqslant \varphi_1(t) \, \| u(t) \|^2 + \cdots,$$

where $A(t)$ is a linear operator. See

G. Ladas and V. Lakshmikantham, in *Ordinary Differential Equations*, 1971 NRL-MRC Conf., L. Weiss (ed.), Academic Press, New York, 1972, pp. 473–487,

where reference to work of Agmon, Nirenberg and others may be found.

See also

U. Mosco, *Convergence of Solutions of Variational Inequalities, Theory and Application of Monotone Operators*, Proc. NATO Advanced Study Inst., Venice, Italy, June, 1968,

as well as other articles in this volume.

§10.24

R. Bellman and K. Cooke, "Existence and Uniqueness Theorems in Invariant Imbedding—II: Convergence of a New Difference Algorithm," *J. Math. Anal. Appl.*, Vol. 12, 1965, pp. 247–253.

Chapter 11 QUASILINEARIZATION

11.1. Introduction

In this chapter, we wish to pursue another analytic approach to the problem of obtaining approximate solutions to nonlinear differential equations. Our aim is to derive upper or lower bounds for the solutions and even a representation for the solution in terms of a maximum or minimum operation.

The basic idea is a simple one. Suppose that the equation has the form

$$\frac{du}{dt} = h(u), \tag{11.1.1}$$

subject, say, to the initial condition $u(0) = c$. Suppose further that we can write

$$h(u) = \max_v g(u, v) \tag{11.1.2}$$

so that (11.1.1) takes the form

$$\frac{du}{dt} = \max_v g(u, v). \tag{11.1.3}$$

It follows that

$$\frac{du}{dt} \geqslant g(u, v) \tag{11.1.4}$$

for any function v. As we have seen in the previous chapter, in certain favorable circumstances this implies that

$$u \geqslant w(v, t), \tag{11.1.5}$$

where $w(v, t)$ is the solution of the corresponding equation

$$\frac{dw}{dt} = g(w, v), \qquad w(0) = c, \tag{11.1.6}$$

with v an as yet unspecified function. Choosing v in an adroit fashion, we can obtain a useful lower bound for u in this way. The most important application of this technique, of course, is that where $g(w, v)$ is linear in w. We, therefore, call this general procedure "quasilinearization." We shall pursue this idea for some particular equations, including the Riccati equation, and then show that it can be made the basis of a method of successive approximations with two qualities, monotonicity and rapid convergence. We shall also consider methods for obtaining upper bounds along the same lines. The methods will then be applied to the matrix Riccati equation and to the second-order differential equation subject to a two-point boundary condition. Briefly, at the close of the chapter, we will indicate the extension of these methods to certain classes of partial differential equations. References to detailed investigations will be found in the Bibliography.

In the chapter following, devoted to dynamic programming, the genesis of these techniques will be described, and interpretations will be given to what appear here to be merely analytic devices.

11.2. The Riccati Equation

Let us begin with the Riccati equation

$$\frac{du}{dt} = u^2 + a(t), \qquad u(0) = c. \tag{11.2.1}$$

As we know, we can write

$$u^2 = \max_v [2uv - v^2], \tag{11.2.2}$$

and thus (11.2.1) may be written in the quasilinear form

$$\frac{du}{dt} = \max_v [2uv - v^2 + a(t)], \qquad u(0) = c. \tag{11.2.3}$$

The maximum operation implies that we have

$$\frac{du}{dt} \geqslant 2uv - v^2 + a(t), \qquad u(0) = c, \tag{11.2.4}$$

for any function $v(t)$. Using the result of Sec. 10.3, we can then assert that for *any* function v we have the inequality

$$u \geqslant w(v, t), \tag{11.2.5}$$

where $w(v, t)$ is the readily determined solution of

$$\frac{dw}{dt} = 2wv - v^2 + a(t), \qquad w(0) = c. \tag{11.2.6}$$

Specifically,

$$w(v, t) = \left[c \exp\left(2\int_0^t v\, ds\right) + \int_0^t \exp\left(2\int_{t_1}^t v\, ds\right) [a(t_1) - v^2(t_1)]\, dt_1\right]. \quad (11.2.7)$$

11.3. Explicit Representation

Returning to (11.2.5), we see that we have established that $u \geqslant w(v, t)$ for all v. Let us note furthermore that equality holds for one function, $v = u$. Hence, we can write

$$u = \max_v w(v, t). \quad (11.3.1)$$

This furnishes an explicit analytic representation for the Riccati equation in terms of the elementary operations of analysis and the maximum operation. Its value lies not so much in the explicit representation as in the pointwise estimates that can be obtained from (11.2.5) by suitable choice of $v(t)$.

11.4. Successive Approximations and Monotone Convergence

Observe that when we write

$$u' = \max_v [2uv - v^2 + a(t)], \qquad u(0) = c, \quad (11.4.1)$$

the function which maximizes on the right is u itself. This strongly suggests that we employ a method of successive approximations in which v is chosen at each stage to be an estimate of the desired solution u.

Let u_0 be an initial approximation to u and determine u_1 by the equation

$$u_1' = 2u_1u_0 - u_0^2 + a(t), \qquad u_1(0) = c. \quad (11.4.2)$$

Having obtained u_1 in this fashion, let u_2 be determined by the equation

$$u_2' = 2u_2u_1 - u_1^2 + a(t), \qquad u_2(0) = c. \quad (11.4.3)$$

Continuing in this fashion, we construct the sequence of functions $\{u_n\}$ where

$$u_{n+1}' = 2u_{n+1}u_n - u_n^2 + a(t), \qquad u_{n+1}(0) = c. \quad (11.4.4)$$

Let us now establish the fundamental result that the sequence thus generated is monotone increasing,

$$u_1 \leqslant u_2 \leqslant \cdots \leqslant u_n \leqslant u_{n+1} \leqslant \cdots, \quad (11.4.5)$$

and, furthermore, that a common bound for the sequence is $u(t)$,

$$u_1 \leqslant u_2 \leqslant \cdots \leqslant u_n \leqslant u_{n+1} \leqslant \cdots \leqslant u(t). \tag{11.4.6}$$

The interval within which this holds will be specified below.

To begin with, let us demonstrate that

$$u_n(t) \leqslant u(t) \tag{11.4.7}$$

for $n = 1, 2, \ldots$. This is important since it shows that the sequence $\{u_n\}$ is well-defined in any interval $[0, t_0]$ within which the solution $u(t)$ exists. To demonstrate (11.4.7), we observe that for any n,

$$2u_n u_{n+1} - u_{n+1}^2 \leqslant 2u_n u_n - u_n^{\,2}. \tag{11.4.8}$$

(since $2u_n u_{n+1} \leqslant u_n^{\,2}\, u_{n+1}^2$). Hence,

$$u_n' \leqslant 2u_n u_n - u_n^{\,2} + a(t) = u_n^{\,2} + a(t), \qquad u_n(0) = c. \tag{11.4.9}$$

Thus, we suspect that u_n is majorized by the solution of the equality, namely the function $u(t)$. However, we have not yet established a majorization result of this type. Hence, we proceed as follows. We have for any n

$$u^2 \geqslant 2uu_n - u_n^{\,2}. \tag{11.4.10}$$

Thus,

$$u' = u^2 + a(t) \geqslant 2uu_n - u_n^{\,2} + a(t), \qquad u(0) = c. \tag{11.4.11}$$

Comparing (11.4.11) with (11.4.4), the defining equation for u_{n+1}, we see from Sec. 10.3 that $u \geqslant u_{n+1}$ within the interval of existence of $u(t)$. The point is that we do possess a majorization theorem for linear equations.

To show that (11.4.5) holds, we note that

$$u_n' = 2u_n u_{n-1} - u_{n-1}^2 + a(t) \leqslant 2u_n u_n - u_n^{\,2} + a(t). \tag{11.4.12}$$

Comparing this inequality for u_n with the equation

$$u_{n+1}' = 2u_{n+1} u_n - u_n^{\,2} + a(t), \tag{11.4.13}$$

we see that we have the desired inequality connecting u_n and u_{n+1}.

It follows from the monotonicity and the uniform boundedness that the sequence $\{u_n(t)\}$ converges for $0 \leqslant t \leqslant t_0$.

Exercises

1. Consider the equation $u' = u^2 + (1 + \epsilon b(t)$, $u(0) = c$, and determine the first two functions u_1, u_2 to powers of ϵ^3 where $u_0 = c$.
2. Carry out the same calculations for $u' = -u^2 + (1 + \epsilon b(t))$.
3. Consider the equation $u' = -u^2 + t$, $u(0) = 1$. Find the power series representations up to t^4 of u_1 and u_2, and compare with the corresponding approximations obtained from $u_0 = 1$, $u_1' = u_0{}^2 + t$, $u_1(0) = 1$, $u_2' = -u_1{}^2 + t$, $u_2(0) = 1$.
4. Carry out the same calculations for $u' = -u^2 + tu$, $u(0) = 1$.

11.5. Maximum Interval of Convergence

What is rather interesting about the foregoing result is that it shows that $\{u_n(t)\}$ possesses the maximum interval of convergence of any sequence constructed by means of successive approximations. Whatever the interval of existence of the solution $u(t)$ of (11.2.1) this is the interval of convergence of $\{u_n(t)\}$.

11.6. Dini's Theorem and Uniform Convergence

A basic result of Dini is that a sequence of continuous functions converging monotonically to a continuous function $u(t)$ in a closed interval $[0, t_0]$ must converge uniformly in this interval. Applying this to the foregoing case, we can assert that the sequence $\{u_n\}$ defined above converges uniformly as well as monotonically.

Exercise

1. Considering the sequence $\{t^n\}$ for $0 \leqslant t \leqslant 1$, show that the condition that the limit function be continuous is essential.
2. Establish Dini's theorem.

11.7. Newton–Raphson–Kantorovich Approximation

Given the scalar equation

$$u' = g(u, t), \qquad u(0) = c, \tag{11.7.1}$$

the usual Picard method of successive approximations leads to the sequence $\{u_n\}$ defined by

$$u'_{n+1} = g(u_n, t), \qquad u_{n+1}(0) = c, \tag{11.7.2}$$

$n \geqslant 0$, with u_0 determined say by the condition $u_0 = c$.

The Newton–Raphson method, suitably extended to the case of functional equations, yields the sequence $\{u_n\}$ defined by

$$u'_{n+1} = g(u_n, t) + (u_{n+1} - u_n)\, g_u(u_n, t), \qquad u_{n+1}(0) = c, \tag{11.7.3}$$

$n \geqslant 0$. Since the extensive study of this procedure for functional equations was first carried out by Kantorovich, the method in this context is usually called the Newton–Raphson–Kantorovich method. For the special case of $g(u, t) = u^2 + a(t)$, reduces to (11.4.4). As we see, the method of quasilinearization yields the Newton–Raphson–Kantorovich procedure if we successively choose v equal to the last approximation obtained. In general, more efficient prediction techniques can be used to obtain an estimate of u_n given $u_1, u_2, \ldots, u_{n-1}$, particularly when we are far from the exact solution. We shall return to this point in Chapter 12 where the meaning of the method of successive approximations used here is examined.

11.8. Quadratic Convergence

What is to be expected in view of the connection with the Newton–Raphson method for the solution of polynomial equations (and quite important in many applications where time of computation is a factor) is that the convergence of $\{u_n\}$ is quadratic, namely that

$$\max_t |u - u_n| \leqslant k_1 (\max_t |u - u_{n-1}|)^2, \tag{11.8.1}$$

for some constant k_1, where the maximum with respect to t is taken over $[0, t_0]$. Let us note in passing that rapid convergence of an algorithm is also desirable to avoid the buildup of round-off error.

To show this quadratic convergence in the case of the Riccati equation, we write

$$\begin{aligned} u_n' &= u_{n-1}^2 + 2(u_n - u_{n-1})\, u_{n-1} + a(t), & u_n(0) &= c, \\ u' &= u_{n-1}^2 + 2(u - u_{n-1})\, u_{n-1} + (u - u_{n-1})^2 + a(t), & u(0) &= c. \end{aligned} \tag{11.8.2}$$

econd equation is a consequence of the identity

$$u^2 = (u_{n-1} + u - u_{n-1})^2 = u_{n-1}^2 + 2u_{n-1}(u - u_{n-1}) + (u - u_{n-1})^2. \quad (11.8.3)$$

It follows that

$$(u - u_n)' = 2(u - u_n)\, u_{n-1} + (u - u_{n-1})^2. \quad (11.8.4)$$

Hence, regarding this as a linear inhomogeneous equation for $(u - u_n)$ with the forcing term $(u - u_{n-1})^2$, we have

$$(u - u_n) = \exp\left(2 \int_0^t u_{n-1}\, ds\right) \int_0^t \exp\left(-2 \int_0^{t_1} u_{n-1}\, ds\right) (u - u_{n-1})^2\, dt_1\,, \quad (11.8.5)$$

and thus

$$\begin{aligned} | u - u_n | &\leqslant \int_0^t \exp\left(2 \int_{t_1}^t u\, ds\right) (u - u_{n-1})^2\, dt_1 \\ &\leqslant \left(\int_0^t e^{2k(t-t_1)}\, dt_1\right) [\max_t | u - u_{n-1} |]^2, \end{aligned} \quad (11.8.6)$$

where $k = \max_t | u |$. Thus, we obtain the inequality

$$\begin{aligned} \max_t | u - u_n | &\leqslant \max_t \left[\int_0^t e^{2k(t-t_1)}\, dt_1\right] \max_t (u - u_{n-1})^2 \\ &\leqslant k_1 (\max_t | u - u_{n-1} |)^2, \end{aligned} \quad (11.8.7)$$

the desired result.

11.9. Upper Bounds

It is not as easy to obtain upper bounds. Suppose, however, that we can find a function w such that

$$\frac{dw}{dt} \geqslant w^2 + a(t), \qquad w(0) = c. \quad (11.9.1)$$

Then we can assert that $w \geqslant u$. To demonstrate this, observe that (11.9.1) implies

$$\frac{dw}{dt} \geqslant w^2 + a(t) \geqslant 2wu - u^2 + a(t), \quad (11.9.2)$$

while u satisfies the equation

$$\frac{du}{dt} = 2uu - u^2 + a(t). \quad (11.9.3)$$

Hence, by the comparison theorem for linear equations we have $w \geqslant u$.

Exercise

1. How would one find functions satisfying (11.9.1) in some initial interval $[0, t_0]$?

11.10. $u' = g(u, t)$

Completely analogous results may be obtained for the equation

$$u' = g(u, t), \qquad u(0) = c, \tag{11.10.1}$$

under the assumption that $g(u, t)$ is uniformly convex in u in some t-interval $[0, t_0]$. We start with the equation (see (9.7.4)).

$$u' = \max_v [g(v, t) + (u - v)\, g_v(v, t)], \tag{11.10.2}$$

and proceed as above to obtain an explicit representation, establish monotonicity and bounded convergence, and finally quadratic convergence.

Exercise

1. Obtain a representation for the solution of $u' = u^k + g(t)$, $u(0) = c$, $k > 1$, in the foregoing fashion.
2. Obtain a representation for the solution of $u' = e^{bu} + g(t)$, $u(0) = c$, in an analogous fashion.
3. Obtain analogous results in the case where $g(u, t)$ is a strictly concave function, i.e., $g_{uu} < 0$, beginning with the representation $g(u, t) = \min_v [g(v, t) + (u - v)g_v(v, t)]$.
4. Consider the vector equation $x' = g(x. t)$, $x(0) = c$, where $g(x, t)$ is a strictly convex function of x for $0 \leqslant t \leqslant t_0$. Under the assumption that $\partial g_i / \partial x_j \geqslant 0$, $i \neq j$, obtain lower bounds for x in terms of the solution of an associated linear equation.

11.11. Random Equation

Consider next an equation of the form

$$u' = g(u, r(t)), \qquad u(0) = c, \tag{11.11.1}$$

where $r(t)$ is a random process, and c possibly a random variable. If g is uniformly convex, we have, as above,

$$u' \geqslant g(v, r(t)) + (u - v)\, g_v(v, r(t)), \tag{11.11.2}$$

for any random process v and value of c. Hence,

$$u \geqslant w \tag{11.11.3}$$

for all such v and c where w is the solution of the corresponding equality. Thus, for any value z,

$$\text{Prob}(u \geqslant z) \geqslant \text{Prob}(w \geqslant z), \tag{11.11.4}$$

where the probability is calculated with respect to the random process $r(t)$ and the random variable c with equality if $v = u$. Hence, we may write

$$\text{Prob}(u \geqslant z) = \max_v[\text{Prob}(w \geqslant z)]. \tag{11.11.5}$$

This can be used to study the behavior of solutions of

$$w'' + (a^2 + r(t))w = 0 \tag{11.11.6}$$

where $r(t)$ is a random process. Questions of this nature are of importance in connection with wave propagation through random media.

11.12. Upper and Lower Bounds

As mentioned above, it is not easy to obtain upper and lower sequences bounding the solution in general. In certain important cases, however, we can do so by means of a simple artifice. To illustrate, consider the equation

$$u' = a(t) - u^2, \qquad u(0) = c, \tag{11.12.1}$$

associated with the second-order linear differential equation

$$w'' - a(t)w = 0. \tag{11.12.2}$$

We consider the frequently occurring case where $a(t)$ is positive. Using the representation for u^2 we can obtain lower bounds for u in the fashion previously described. Making the change of variable $u = z^{-1}$, we derive the equation

$$z' = 1 - a(t)\, z^2, \tag{11.12.2}$$

to which the same procedure can be applied by virtue of the positivity of $a(t)$. The point of this is that a lower bound for $z = u^{-1}$ automatically furnishes an upper bound for u. We shall present the details of an application of this artifice below.

Exercise

1. In what interval does this device work for $u' = u^2 + a(t)$ where $a(t) > 0$?

11.13. Asymptotic Behavior

Let us now use the foregoing results to deduce the asymptotic behavior of solutions of

$$u'' - (1 + f(t))u = 0, \tag{11.13.1}$$

under the assumption that $f(t) \to 0$ as $t \to \infty$.

As we have seen in Chapter 1, Volume I, under reasonable assumptions concerning $a(t)$, the more general equation

$$u'' - a(t)u = 0 \tag{11.13.2}$$

can be transformed into the foregoing form. Hence, it is useful to study the particular equation in (11.13.1).

Let us show by means of the foregoing techniques that there is a solution u such that as $t \to \infty$,

$$v = \frac{u'}{u} \to 1. \tag{11.13.3}$$

We have already established this using classical techniques in Chapter 1, Volume I.

The equation for v is

$$v' = 1 + f(t) - v^2, \qquad v(0) = v_0, \tag{11.13.4}$$

a Riccati equation. We know from what has preceded that we can write

$$v = \min_w F(w, t), \tag{11.13.5}$$

where

$$F(w, t) = \left[v_0 \exp\left(-2\int_0^t w\, dt_1\right) + \int_0^t \exp\left(-2\int_s^t w\, dt_1\right)[w^2 + (1 + f(s))]\, ds\right]. \tag{11.13.6}$$

Let us use the initial approximation $w = 1$, a reasonable approximation to the solution of (11.13.4) when $f(t)$ is small.

We thus obtain, using (11.13.6), the inequality

$$\begin{aligned} v &\leqslant \left[v_0 e^{-2t} + e^{-2t} \int_0^t e^{2t_1}[2 + f(t_1)]\, dt_1\right] \\ &\leqslant (1 - e^{-2t}) + v_0 e^{-2t} + e^{-2t} \int_0^t e^{2t_1} f(t_1)\, dt_1 \\ &\leqslant 1 + o(1) \end{aligned} \tag{11.13.7}$$

as $t \to \infty$. The condition $f(t) \to 0$ as $t \to \infty$ implies that

$$e^{-2t} \int_0^t e^{2t_1} f(t_1)\, dt_1 \to 0 \tag{11.13.8}$$

as $t \to \infty$.

To obtain an inequality in the other direction, we use the change of variable $u = v^{-1}$ which, as explained above, yields the result

$$u \geqslant \frac{1}{G(z, t)}, \tag{11.13.9}$$

where

$$G(z, t) = \left[v_0^{-1} \exp\left(-2 \int_0^t (1 + f(t_1))\, z(t_1)\, dt_1\right) + \int_0^t \exp\left(-2 \int_s^t (1 + f(t_1))\, z(t_1)\, dt_1\right) [1 + (1 + f(s))\, z^2]\, ds\right]. \tag{11.13.10}$$

Once again, use $z = 1$ as an initial approximation. A straightforward estimation yields

$$G(1, t) = 1 + o(1) \tag{11.13.11}$$

as $t \to \infty$. Comparison of (11.13.7) and (11.13.11) yields the desired asymptotic result.

Exercises

1. Show directly from the differential equation that $v \to 1$ as $t \to \infty$ if v_0 is large and positive. What happens if v_0 is large and negative?

2. Use the initial approximation $w = 1 + f(t)/2$ to calculate improved upper and lower bounds under suitable assumptions concerning $\int^\infty |f'|\, dt$ and $\int^\infty f^2\, dt$.

3. Use the foregoing method to study the asymptotic behavior of the solutions of $u' = \varphi(t) - u^n$, $u' = \varphi(t) - e^{bu}$ under the assumption that $\varphi(t) \to 1$ as $t \to \infty$.

4. Obtain improved upper and lower bounds by means of a change of variable $u = (a(t)v + b(t))/(c(t)v + d(t))$ in the general case $u' = u^2 + f(t)$, where $f(t)$ need not be positive.

11.14. Multidimensional Riccati Equation

Let us now obtain analogues of the foregoing results for the multidimensional case. To simplify the analysis, we will consider the matrix Riccati equation

$$R' = A - R^2, \qquad R(0) = I, \tag{11.14.1}$$

where A is positive definite, rather than the more general

$$R' = A + BR + RC + RDR, \qquad R(0) = I. \tag{11.14.2}$$

In the following chapters we indicate how these matrix Riccati equations arise naturally in classical analysis and in modern control theory.

The basic ideas are the same as before, but some slight complications arise from the fact that positive definite matrices are only partially ordered and from the fact that this ordering has certain idiosyncrasies.

We begin once again with an identity,

$$R^2 = (S + R - S)^2 = S^2 + S(R - S) + (R - S)S + (R - S)^2, \tag{11.14.3}$$

from which we conclude that

$$R^2 \geqslant SR + RS - S^2 \tag{11.14.4}$$

for all symmetric S, with equality if and only if $R = S$. Here inequality is in terms of the partial ordering of nonnegative definiteness as discussed in Sec. 2.21, Volume I. Thus, (11.14.1) leads to the relation

$$R' \leqslant A + S^2 - SR - RS, \qquad R(0) = I. \tag{11.14.5}$$

Consider the solution of the corresponding equation

$$Y' = (A + S^2) - SY - YS, \qquad Y(0) = I, \tag{11.14.6}$$

which we denote by $Y = F(S, t)$. Then the explicit expression for the solution of (11.14.6), given in Sec. 10.3, Exercise 2, allows us to conclude that

$$R \leqslant F(S, t) \tag{11.14.7}$$

for $t \geqslant 0$ and all $S \geqslant 0$.

Next, make the change of variable $R = Z^{-1}$. The equation for Z is then

$$Z' = I - ZAZ, \qquad Z(0) = I. \tag{11.14.8}$$

As above, using a simple identity, we can conclude that

$$ZAZ \geqslant SAZ + ZAS - SAS \tag{11.14.9}$$

for all symmetric S. Hence,

$$Z' \leqslant I + SAS - (SAZ + ZAS), \tag{11.14.10}$$

whence

$$Z \leqslant G(S, t), \tag{11.14.11}$$

where $G(S, t)$ is the solution of

$$W' = I + SAS - (SAW + WAS), \qquad W(0) = I. \tag{11.14.12}$$

Combining (11.14.7) and (11.14.11), we have the desired upper and lower bounds,

$$G(S_1, t)^{-1} \leqslant R \leqslant F(S_2, t), \tag{11.14.13}$$

for $t \geqslant 0$, where $S_1(t)$ and $S_2(t)$ are arbitrary positive definite matrices in a t-interval $[0, t_0]$.

Exercise

1. Obtain analogues of the results of Sec. 11.13 by suitable choice of S_1 and S_2.

11.15. Two-point Boundary Value Problems

Equations of the form

$$u'' = g(u, t), \tag{11.15.1}$$

subject to a two-point boundary condition such as

$$u(0) = c_1, \qquad u(T) = c_2. \tag{11.15.2}$$

arise in a natural fashion in the calculus of variations and in various fields of mathematical physics. Realistic descriptions of physical processes produce multidimensional versions in which a vector form of (11.15.1) appears. Neither the analysis of the structure of the solution, nor its numerical calculation, is routine. One reason for this is that the determination of conditions which ensure existence and uniqueness is itself difficult, as opposed to the situation in the initial-value case. It is easy to construct equations which possess either no solution or an infinite number of solutions.

If $g(u, t)$ is linear in u,

$$g(u, t) = g(t)u + h(t), \tag{11.15.3}$$

existence and uniqueness of the equation may be studied in a direct fashion as we have seen in Chapter One. In the vector case,

$$x'' = A(t)x + b(t), \qquad x(0) = c, \quad x(T) = d, \tag{11.15.4}$$

to carry through the numerical solution it is necessary to combine the ability to resolve initial-value problems with the ability to solve linear systems of algebraic equations.

Assuming that we possess these capabilities, let us examine the application of quasilinearization to problems of this nature. Let us begin with the scalar equation of (11.15.1). Subsequently, we will briefly discuss the vector case.

Consider the sequence $\{u_n\}$ described by the recurrence relation

$$u''_{n+1} = g(u_n, t) + (u_{n+1} - u_n)\, g_u(u_n, t), \qquad u_{n+1}(0) = c_1, \quad u_{n+1}(T) = c_2, \tag{11.15.5}$$

for $n \geqslant 0$, with u_0 specified. For example, u_0 may be taken to be the straight line determined by the two-point boundary condition,

$$u_0 = c_1 + (c_2 - c_1)t/T. \tag{11.15.6}$$

We say "described" since it is not as yet clear that these functions actually exist.

In the following sections we shall discuss various questions connected with the convergence of the sequence $\{u_n\}$, beginning with a demonstration of the fact that (11.15.5) is a meaningful definition of the sequence $\{u_n\}$ under certain assumptions concerning $g(u)$.

11.16. Maximum Interval of Convergence

Let us begin with the case where we have established by some other means the existence of a unique solution of (11.15.1). For example, we may have used an original variational formulation. We further assume that $g(u, t)$ is strictly convex as a function of u for $0 \leqslant t \leqslant T$. Then the equation of (11.15.1) may be written

$$u'' = \max_v [g(v, t) + (u - v)\, g_u(v, t)], \tag{11.16.1}$$

whence

$$u'' \geqslant [g(v, t) + (u - v)\, g_u(v, t)] \tag{11.16.2}$$

for arbitrary v. In particular, we have

$$u'' \geqslant g(u_0\,, t) + (u - u_0)\, g_u(u_0 t) \tag{11.16.3}$$

We wish to compare this inequality with the equation for u_1, namely

$$u_1'' = g(u_0\,, t) + (u_1 - u_0)\, g_u(u_0\,, t), \tag{11.16.4}$$

with u_1 subject to the same boundary conditions,

$$u_1(0) = c_1\,, \qquad u_1(T) = c_2\,. \tag{11.16.5}$$

The comparison hinges upon the properties of the equation

$$w'' - g_u(u_0\,, t)w = f(t), \tag{11.16.6}$$

subject to the boundary conditions $w(0) = w(T) = 0$, since $w = u - u_1$ satisfies this equation with a positive forcing function. If the associated Green's function is nonnegative, we have

$$u - u_1 > 0. \tag{11.16.7}$$

Furthermore, we use this property of the Green's function to demonstrate the fact that u_1 is uniquely determined by (11.15.5). A sufficient condition for this nonnegativity, as we know, is that $g_u > 0$ for all u. See Sec. 10.12. For example,

$$g(u) = u + u^3, \qquad \text{or} \qquad g(u) = e^u, \tag{11.16.8}$$

satisfy this condition. In general, we require a condition such as

$$g_u > -\lambda_1(T), \tag{11.16.9}$$

where $\lambda_1(T)$ is the smallest characteristic value associated with the Sturm–Liouville equation

$$w'' + \lambda w = 0, \qquad w(0) = w(T) = 0. \tag{11.16.10}$$

In this case $\lambda_1(T) = \pi^2/T^2$.

Continuing in this fashion, we demonstrate that

$$u \leqslant u_n \leqslant u_{n-1} \leqslant \cdots \leqslant u_1 . \tag{11.16.11}$$

(See Fig. 11.1.)

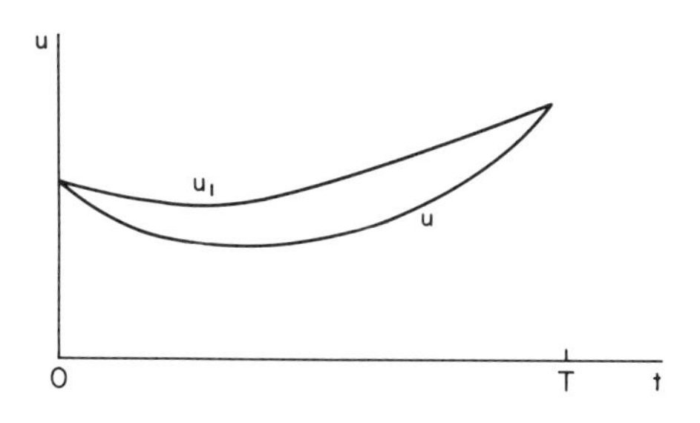

Figure 11.1

In general, where g enjoys no special properties aside from differentiability, the argument is carried through by imposing the condition $T < < 1$. This will ensure that (11.16.9) holds.

In the special, but important, case where $g_u > 0$, $g_{uu} > 0$ the method of successive approximations converges wherever the solution exists. Combining (11.16.11) with the monotonicity of g_u, we see that (11.16.9) holds for each approximant.

11.17. Quadratic Convergence

To establish the convergence of the sequence $\{u_n\}$ from first principles, we proceed as in Sec. 11.8. We convert the recurrence relation of (11.15.5) into the integral equation

$$u_{n+1} = \int_0^T k(t, t_1)[g(u_n , t_1) + (u_{n+1} - u_n) g_u(u_n , t_1)] \, dt_1 , \tag{11.17.1}$$

where $k(t, t_1) \equiv k(t, t_1 , T)$ is the Green's function associated with the simpler linear equation

$$w'' = f, \qquad w(0) = w(T) = 0. \tag{11.17.2}$$

Without any loss of generality, let us take $c_1 = c_2 = 0$, so that (11.17.1) contains no forcing term.

There are now the usual steps in the proof. First we establish inductively that $|u_n|$ is uniformly bounded for $0 \leqslant t \leqslant T$ for $n = 1, 2, \ldots$, provided that T is sufficiently small. Then we establish the convergence of the series $\sum_{n=0}^{\infty} |u_{n+1} - u_n|$. Regarding (11.17.1) as an integral equation for u_{n+1} of the form

$$u_{n+1} = \int_0^T k(t, t_1)[g(u_n, t_1) - u_n g_u(u_n, t_1)]\, dt_1 + \int_0^T k(t, t_1)\, g_u(u_n, t_1)\, u_{n+1}\, dt_1, \qquad (11.17.3)$$

we see that all of this can be made to depend on the magnitude of

$$\int_0^T |k(t, t_1)|\, dt_1 .$$

It is easy to see that we can make this quantity as small as desired by choosing T small. We leave the details of the proof as a series of exercises.

Finally, to establish the desired quadratic convergence, we write, using the mean-value theorem

$$u'' = g(u_n, t) + (u - u_n)\, g_u(u_n, t) + \frac{(u - u_n)^2}{2} g_{uu}(\theta, t), \qquad (11.17.4)$$

where θ lies between u and u_n. Hence, using the equation for u_{n+1} in terms of u_n, we have

$$(u - u_{n+1})'' = g_u(u_n, t)(u - u_{n+1}) + \frac{(u - u_n)^2}{2} g_{uu}(\theta, t). \qquad (11.17.5)$$

Regard this as a linear equation for $u - u_{n+1}$ with the forcing term

$$\frac{(u - u_n)^2}{2} g_{uu}(\theta, t).$$

From this follows

$$\max_t |u - u_{n+1}| \leq k_1 (\max_t |u - u_n|)^2 \qquad (11.17.6)$$

where k_1 is a constant. This is the desired quadratic convergence.

11.18. Discussion

What we have outlined above is a proof of the convergence of the sequence $\{u_n\}$ generated by the Newton–Raphson–Kantorovich method. It is interesting to observe that the sequence $\{u_n\}$ converges quadratically provided only that g_{uu} exists and that the t-interval is sufficiently small. In all important cases, restricting the t-interval will ensure that g_{uu} will be of one sign in $[0, T]$. If it is positive, we use the maximum representation; if negative, the minimum representation

The theory of quasilinearization enables us in a number of important cases to increase the domain of validity of this method of successive approximations while simultaneously giving us a great deal of freedom in obtaining upper and lower bounds.

On the other hand, it is essential to have a tool such as the Newton–Raphson–Kantorovich method which can be employed in very general situations under mild assumptions concerning the structure of the equation and the nature of the boundary conditions.

11.19. Computational Feasibility

Consider the vector equation

$$x'' = g(x, t), \qquad x(0) = c, \quad x(T) = d. \tag{11.19.1}$$

The method of quasilinearization leads to the system of linear equations

$$x''_{n+1} = g(x_n, t) + J(x_n)(x_{n+1} - x_n), \qquad x_{n+1}(0) = c, \quad x_{n+1}(T) = d. \tag{11.19.2}$$

Here $J(x_n)$ is the Jacobian matrix

$$J(x_n) = (\partial g_i(x_n)/\partial x_{nj}), \tag{11.19.3}$$

with x_{nj} the jth component of x_n. Let us avoid the issue of monotonicity, which requires a quite detailed investigation in general, and about which relatively little is known in the multidimensional case, and concentrate on the computational mechanics.

Observe that the foregoing algorithm can be considered a "closure" technique for solving nonlinear equations in terms of the ability to solve linear equations. In order to determine x_{n+1} in terms of x_n, we are required to solve a two-point boundary-value problem. As we have seen in Chapter 2, Volume I, this can be accomplished in terms of the solution of linear differential equations subject to initial conditions, plus the solution of a linear system of algebraic equations. In Chapter 12 we will present a different approach to two-point boundary-value problems which eliminates this last troublesome step at the expense of introducing a different type of difficulty. For the moment, however, let us accept it as part of the solution plan.

There is one further point of great moment which must be mentioned. That is the requirement to store the nth approximant, x_n, which enables the calculation of the $(n+1)$st, x_{n+1}, using (11.19.2). This is of no particular difficulty if the dimension of x is small. If, however, x is of dimension 100 or 500, the demands of rapid-access storage may become exorbitant. The problem of storage has in the past been one of the major hindrances to the widespread use of successive approximations as an effective computational procedure.

Let us now indicate how the device of adjoining, used in Sec. 6.14, Volume I, can be employed here to circumvent the demands of rapid-access storage. The basic idea, as before, is that of replacing the set of values of x_n by an algorithm for calculating these values.

We suppose that x_0 is some function of convenient analytic form so that there are no serious demands for the storage of the values of $x_0(t)$ for $0 \leqslant t \leqslant T$, e.g.,

$$x_0 = c + (d - c)t/T. \tag{11.19.4}$$

Let us turn now to the equation for x_1 in (11.19.2). Carrying out the required operations for the determination of x_1 as the solution of a linear equation subject to a two-point boundary condition, we can obtain the missing initial condition for $x_1'(0)$, say $b^{(1)}$. We then regard x_1 as determined by the solution of the *initial value* problem

$$x_1'' = g(x_0, t) + J(x_0)(x_1 - x_0), \qquad x_1(0) = c, \quad x_1'(0) = b^{(1)}. \tag{11.19.5}$$

Turn now to the equation for x_2,

$$x_2'' = g(x_1, t) + J(x_1)(x_2 - x_1). \tag{11.19.6}$$

To obtain principal and particular solutions of this equation, subject to initial conditions, we solve it simultaneously with (11.19.5). In this way we carry through the process of determining the missing initial condition $x_2'(0)$ without the necessity for storing the function $x_1(t)$ for t in $[0, T]$. Similarly, we solve the equation for x_3 by adjoining the equations for x_1 and x_2, and we continue in this fashion.

As we see, we are relying heavily upon our ability to solve large systems of ordinary differential equations subject to initial conditions. If the procedure outlined above is carried on for too many steps, that is if the dimension of the system becomes too large, eventually we reach the limit of our capacity for this type of operation.

This is where quadratic convergence becomes of great utility. Usually, fewer than ten passes are required. If, by chance, more iterations are required, we can use differential approximation to store the kth approximation, x_k, and then start all over again with x_k as a new initial approximation.

11.20. Elliptic Equations

The application of the quasilinearization procedure to nonlinear partial differential equations of elliptic type such as

$$\begin{aligned} u_{xx} + u_{yy} &= g(u), \qquad (x, y) \in R, \\ u &= h(p), \qquad p \in B, \end{aligned} \tag{11.20.1}$$

where B is the boundary of R, depends upon a knowledge of the majorization properties of the linear operator

$$L(u) = u_{xx} + u_{yy} + q(x, y)u. \tag{11.20.2}$$

We have provided a basis for this in Chapter 10. There is thus no difficulty in obtaining the analogues for the foregoing results, granted a modicum of partial differential equation theory.

Exercise

1. Consider the case where $g(u) = e^u$ in (11.20.1). This is an equation of some interest in magnetohydrodynamics. Assuming the existence and uniqueness of a solution, show that the sequence furnished by quasilinearization converges monotonically and uniformly to this solution for any region R. See also

   J. Keller, "Electrohydrodynamics I: The Equilibrium of a Charged Gas in a Container," *J. Rat. Mech. Anal.*, Vol. 5, 1956, pp. 715–724.

11.21. Parabolic Equations

Similarly, the application of quasilinearization to nonlinear partial differential equations of parabolic type such as

$$\begin{aligned} u_t &= u_{xx} + u_{yy} - g(u), \qquad (x, y) \in R, \quad t > 0, \\ u &= h(p), \qquad p \in B, \\ u(x, y, 0) &= k(x, y), \end{aligned} \tag{11.21.1}$$

depends upon a knowledge of the majorization properties of the linear operator associated with the equation

$$u_t - (u_{xx} + u_{yy} + q(x, y, t)u) = 0. \tag{11.21.2}$$

We have provided a brief indication of how results of this nature may be obtained in Chapter 10.

11.22. Minimum and Maximum Principles

The study of the application of quasilinearization to general classes of partial differential equations brings us into contact with one of the major topics in the theory of partial differential equations, that of "maximum principles." Any discussion would take us too far off course. Consequently, we shall refer the interested reader to the bibliography at the end of the chapter.

Exercises

1. Show that the conservation law $u_t + f(u)_x = 0$, $u(x, 0) = g(x)$ may be written in the form $w_t + f(w_x) = 0$ with a suitable change of variable.

2. Consider the associated linear equation

$$z_t + f(v) + (z_x - v)f'(v) = 0, \qquad z(x, 0) = g_1(x).$$

   Write $z = z(x, t, v)$. Show that $z = \min_v w(x, t, v)$.

3. Show that we may restrict v to be constant in the minimization, and thus derive the Hopf–Lax representation formula. (The argument follows P. Lax. The details may be found on page 567 of the paper by R. Kalaba cited below. For the original results see

   E. Hopf, "The Partial Differential Equation $u_t + uu_x = \mu u_{xx}$," *Comm. Pure Appl. Math.*, Vol. 3, 1950, pp. 201–230.

   P. Lax, *Initial Value Problems for Nonlinear Hyperbolic Equations*, Univ. of Kansas, Tech. Rep. No. 14, 1955.

Miscellaneous Exercises

1. Let the x_i be real and nonnegative and $p > 1$. Let $q = p/p - 1$. Then

$$M_p(x) = \left(\sum_{i=1}^{N} x_i^p\right)^{1/p} = \max_y \sum_{i=1}^{N} x_i y_i ,$$

where the y_i range over $y_i \geqslant 0$, $\sum_{i=1}^N y_i^q \leqslant 1$.

2. Hence, establish Minkowski's inequality,

$$M_p(x + y) \leqslant M_p(x) + M_p(y),$$

where the components of x and y are nonnegative. For further applications of this technique to the theory of inequalities and discussion, see

E. F. Beckenbach and R. Bellman, *Inequalities*, Springer–Verlag, Berlin and New York, 1961.

3. Consider the equation $r + r^n = a$, $n > 1$, $a > 0$. Starting with the expression

$$r^n = \max_{s \geqslant 0}[nrs^{n-1} - (n - 1)\, s^n],$$

show that

$$r = \min_s \left[\frac{a + (n - 1)\, s^n}{1 + ns^{n-1}}\right].$$

4. Perform the change of variable $r^n = t$, obtaining $t^{1/n} + t = a$ and thus derive a representation for t and thus r in terms of maximum operation.

5. Consider the equation $g(u) = 0$, where g is strictly convex. From the representation

$$g(u) = \max_v [g(v) + (u - v)\, g'(v)],$$

show that we can write the solution in the form

$$u = \min_v \left[v - \frac{g(v)}{g'(v)}\right].$$

6. Consider the problem of minimizing the functional

$$J(u) = \int_0^T g(u, u')\, dt$$

over some class of functions. Show that quasilinearization applied to the Euler equation is equivalent to obtaining the Euler equation of

$$J(u, v) = \int_0^T g_2(u, u', v, v')\, dt,$$

where $g_2(u, u', v, v')$ consists of the first three terms of $g(u, u')$ around v, v'. For the convergence of this method of successive approximations, see

E. S. Levitin and B. T. Poljak, "Convergence of Minimizing Sequences in Conditional Extremum Problems," *Soviet Math.*, Vol. 7, 1966, pp. 764–767.

7. Let $g(u)$ be strictly convex and let $g_1(u)$ be obtained from $g(u)$ by means of a polygonal approximation from above. Let v be the solution of $v' = g_1(v)$, $v(0) = c$, u that of $u' = g(u)$, $u(0) = 0$. Do we have a relation of the form $u \leqslant v$?

8. Consider the corresponding question for $u'' = g(u)$, $u(0) = c_1$, $u(T) = c_2$. (In the following chapter, we will consider a procedure for obtaining accurate polygonal approximations).

9. Consider the differential equation $u' = -g(u)/g'(u)$, $u(0) = c$. Under what conditions does $u(t)$ possess a limit as $t \to \infty$ and converge to a solution of $g(u) = 0$?

10. Show that $g(u) = g(c)e^{-t}$ if the solution exists for all $t > 0$.

11. Similarly, consider the equation $x' = -J(g)^{-1}g$, $x(0) = c$ as a method for obtaining solutions of

$$g(x) = \begin{pmatrix} g_1(x_1, x_2) \\ g_2(x_1, x_2) \end{pmatrix} = 0.$$

Show that $g(x) = e^{-t}g(c)$. Here $J(g)$ is the Jacobian matrix of g, i.e.,

$$J(g) = \begin{pmatrix} \partial g_1/\partial x_1 & \cdots \\ \cdots & \cdots \end{pmatrix}.$$

Bibliography and Comments

§11.1. The theory of quasilinearization was first presented in

R. Bellman, "Functional Equations in the Theory of Dynamic Programming—V: Positivity and Quasilinearity," *Proc. Nat. Acad. Sci. USA*, Vol. 41, 1955, pp. 743–746.

It was then developed in

R. Kalaba, "On Nonlinear Differential Equations, The Maximum Operation, and Monotone Convergence," *J. Math. Mech.*, Vol. 8, 1959, pp. 519–574.

For more detailed results, see

R. Bellman, and R. Kalaba, *Quasilinearization and Nonlinear Boundary-value Problems*, American Elsevier, New York, 1965.

The explicit representation of the Riccati equation has been used to study a number of questions in scattering theory; see

F. Calogero, "A Novel Approach to Elementary Scattering Theory, *Nuovo Cimento*, Vol. 27, 1963, pp. 261–302.

F. Calogero, "A Variational Principle for Scattering Phase Shifts," *Nuovo Cimento*, Vol. 1963, pp. 947–951.

F. Calogero, "A Note on the Riccati Equation," *J. Math. Phys.*, Vol. 4, 1963, pp. 427–430.

F. Calogero, "Maximum and Minimum Principles in Potential Scattering," *Nuovo Cimento*, Vol. 28, 1963, pp. 320–333.

F. Calogero, *Variable Phase Approach to Potential Scattering*, Academic Press, New York, 1967.

§11.6. There are extensive generalizations of Dini's theorem; see

K. Vala, "On Compact Sets of Compact Operators," *Ann. Acad. Sci. Fonn.*, Ser. A, Vol. 351, 1946, pp. 3–8.

W. Mlak, "Note on Abstract Differential Inequalities and Chaplighin Method," *Ann. Polon. Math.*, Vol. 10, 1961, pp. 253ff.

W. Mlak, "Note on Maximum Solutions of Differential Equations," *Contributions Differential Equations*, Vol. 1, 1963, pp. 461–465.

S. Karlin, "Positive Operators," *J. Math. Mech.*, Vol. 8, 1959, pp. 907–937 (esp. p. 911).

§11.7. For a detailed discussion of why Raphson's name should be given prominence, see

D. T. Whiteside, "Patterns of Mathematical Thought in the Latter Seventeenth Century," *Arch. Hist. Exact Sci.*, Vol. 1, 1961, pp. 179–388 (esp. p. 207).

Some important books and papers are

L. V. Kantorovich, "Functional Analysis and Applied Mathematics," *Uspehi Mat. Nauk*, Vol. 3, 1948, pp. 89–185.

M. M. Vainberg, *Variational Methods for the Study of Nonlinear Operators*, Holden-Day, San Francisco, California, 1964.

J. M. Ortega, and W. C. Rheinboldt, "Monotone Iterations for Nonlinear Equations with Applications to Gauss-Seidel," *SIAM J. Numer. Anal.*, Vol. 4, 1967, pp. 171–190.

J. T. Vandergrift, "Newton's Method for Convex Operators in Partially Ordered Spaces," *SIAM J. Numer. Anal.* Vol., 4, 1967, pp. 406–432.

L. F. Shampine, "Monotone Iterations and Two-sided Convergence," *SIAM J. Numer. Anal.*, Vol. 3, 1966, pp. 607–615.

L. Collatz, "Monotonie und Extremalprinzipien beim Newtonsche Verfahren," *Numer. Math.*, Vol. 3, 1961, pp. 99–106.

L. Collatz, *Funktionalanalysis und Numerische Mathematik*, Springer–Verlag, Berlin, 1964; English translation, Academic Press, New York, 1966.

§11.8. For an interesting use of second-order approximation techniques, see

J. Moser, "A New Technique for the Construction of Solutions of Nonlinear Differential Equations," *Proc. Nat. Acad. Sci. USA*, Vol. 47, 1961, pp. 1824–1831.

§11.10. See

R. Bellman, "On Monotone Convergence to Solutions of $u' = g(u, t)$," *Proc. Amer. Math. Soc.*, Vol. 8, 1957, pp. 1007–1009.

§11.11. See

R. Bellman, "On the Representation of the Solution of a Class of Stochastic Differential Equations," *Proc. Amer. Math. Soc.*, Vol. 9, 1958, pp. 326–327.

R. Bellman, T. T. Soong, and R. Vasudevan, "Quasilinearization and Stability of Nonlinear Stochastic Systems, *J. Math. Anal. Appl.* (to appear).

§11.13. For an interesting alternate approach, see

R. Wilcox, "Bounds for Approximate Solutions to Operator Differential Equations," *J. Math. Anal. Appl.* (to appear).

§11.15. See the book by Bellman and Kalaba cited above for further discussion, examples, and references to a number of applications. See also

W. J. Coles and T. L. Sherman, "Convergence of Successive Approximations for Nonlinear Two-point Boundary-value Problems," *SIAM J. Appl. Math.*, Vol. 15, 1967, pp. 426–433.

§§11.20–11.21. Numerical results will be found in the Bellman–Kalaba volume previously cited.

§11.22. See the Beckenbach–Bellman book previously cited for a number of references.

Chapter 12 DYNAMIC PROGRAMMING

12.1. Introduction

In this chapter, we wish to present some of the fundamental ideas and techniques of the theory of dynamic programming, a theory whose origin lies in the domain of multistage decision processes. With the usual mathematical license, however, we extend it to cover other problem areas which can profitably be interpreted as multistage decision processes.

We have limited objectives at the moment. We wish first to show how the fundamental concept of "approximation in policy space" leads quite naturally to the theory of quasilinearization and motivates our interest in monotone operators. Then we wish to present a new approach to the calculus of variations as a multistage decision process of continuous deterministic type which leads both to a new treatment of two-point boundary-value problems and to new methods for handling partial differential equations of the form

$$u_t = g(u, u_x), \qquad u(x, 0) = h(x). \tag{12.1.1}$$

Analogously, stochastic control processes provide new insight into the equation

$$u_t = u_{xx} + g(u, u_x). \tag{12.1.2}$$

Many interesting new classes of analytic and computational problems arise in this fashion.

A characteristic of the dynamic programming approach is the avoidance of solving systems of linear algebraic equations in connection with two-point boundary-value problems of linear type and the replacement of this task by the solution of matrix Riccati equations. We shall, however, briefly discuss the use of dynamic programming to solve ill-conditioned systems, using a method of Tychonoff as a starting point. Finally we present an application to adaptive polygonal approximation and then to spline approximations.

In the exercises we shall indicate many novel types of nonlinear equations that arise in the theory of dynamic programming.

12.2. Multistage Processes

We begin with the following abstraction of a physical process. We suppose that a system S is described by a finite-dimensional vector p, the state vector, where $p \in R$, a specified domain, and that at discrete times, 0, 1,..., a fixed transformation $T(p)$ is applied; it is assumed that $T(p) \in R$, whenever $p \in R$. In this way, we generate a sequence of states, $\{p_n\}$, where

$$p_{n+1} = T(p), \qquad n = 0, 1, 2, \ldots . \tag{12.2.1}$$

This is the genesis of the theory of iteration, a subject we will treat in some detail in Chapter 14. If we set

$$p_n = f_n(p), \qquad n = 0, 1, 2, \ldots, \tag{12.2.2}$$

with $f_0(p) = p$, an acknowledgment that the state at time n depends upon the initial state, we see that this function satisfies the relations

$$f_{n+1}(p) = f(f_n(p)) = f_n(f(p)). \tag{12.2.3}$$

Generally, an inductive argument shows that

$$f_{m+n}(p) = f_m(f_n(p)), \tag{12.2.4}$$

for $m, n = 0, 1, \ldots$, a basic semigroup property.

12.3. Continuous Version

A continuous version of the foregoing discrete process is furnished by a system described by the vector $x(t)$ where x satisfies the finite-dimensional differential equation

$$\frac{dx}{dt} = g(x), \qquad x(0) = c. \tag{12.3.1}$$

If we set

$$x(t) = f(c, t), \tag{12.3.2}$$

acknowledgment of the fact that the state at time t depends upon the initial state c, we see that uniqueness of the solution of (12.3.1) (assumed), leads to the basic semigroup relation

$$f(c, s + t) = f(f(c, s), t), \tag{12.3.3}$$

for $s, t \geqslant 0$, with $f(c, 0) = c$. This relation too will be explored in greater detail in Chapter 14. In the meantime, we wish merely to set the stage for multistage decision processes.

Exercises

1. Using (12.3.3) and the differential equation $u' = u$, show that $e^{s+t} = e^s e^t$.
2. Using (12.3.3) and $u'' + u = 0$, derive the addition formulas for $\sin t$ and $\cos t$.

12.4. Multistage Decision Processes

Let us now suppose that in place of a single transformation $T(p)$ at each stage, we have available a family of transformations $T(p, q)$, where $p \in R$ as before and q belongs to a space D. The vector q is called the decision vector, and D a decision space. At time 0, we choose a particular p, q_1, and produce the point

$$p_1 = T(p, q_1). \tag{12.4.1}$$

We assume that $p_1 \in R$ whenever $p \in R$ and $q_1 \in D$. At time 1, we choose $q = q_2$ and produce the new point

$$p_2 = T(p_1, q_2). \tag{12.4.2}$$

Continuing in this fashion, a sequence of choices (or decisions) q_1, $q_2, \ldots, q_N$ produces a sequence of states $p_1, p_2, \ldots, p_N$, where

$$p_{n+1} = T(p_n, q_{n+1}), \qquad n = 0, 1, \ldots, N-1. \tag{12.4.3}$$

See Fig. 12.1. This yields a trajectory in phase space.

Let

$$R_N = R(p, p_1, p_2, \ldots, p_N; q_1, q_2, \ldots, q_N) \tag{12.4.4}$$

be a prescribed scalar function of the succession of state vectors p_i and decision vectors q_i, $i = 1, 2, \ldots, N$. If the q_i are chosen so as to maximize

$$\overset{q_1}{\underset{p}{\bullet\!\!\longrightarrow}} \quad \overset{q_2}{\underset{p_1}{\bullet\!\!\longrightarrow}} \quad \overset{q_3}{\underset{p_2}{\bullet\!\!\longrightarrow}} \quad \cdots \quad \overset{q_{n+1}}{\underset{p_n}{\bullet\!\!\longrightarrow}}$$

Figure 12.1

R_N, we call this process a *multistage decision process*. The mathematical theory constructed to treat processes of this general nature is called dynamic programming.

12.5. Stochastic and Adaptive Processes

The foregoing description of a multistage decision process of deterministic type can readily be extended to cover processes where there are features of uncertainty. Thus, we can consider the case where the transformation has the form $T(p, q, r)$ with r a random variable with a known distribution function, and then continue from there to treat many interesting and quite important processes involving various types of partial information. References to extensive investigations of questions of this nature will be found at the end of the chapter. What is rather remarkable is that many of these processes shed considerable light on deterministic aspects of the theory of partial differential equations. We shall comment briefly on this in Sec. 12.25.

Exercises

1. Show that the maximum operation possesses the following property: $\max[a_1, a_2, \ldots, a_N] = \max[a_1, \max[a_2, \ldots, a_N]]$.
2. Similarly, show that $\min[a_1, a_2, \ldots, a_N] = \min[a_1, \min[a_2, \ldots, a_N]]$.
3. If $M(a_1, a_2, \ldots, a_N) = M(a_1, M(a_2, \ldots, a_N))$, $a_i \geqslant 0$, $N = 1, 2, \ldots$, what form must M have?

   See

   J. Aczel, *Lectures on Functional Equations and their Applications*, Academic Press, New York, 1966.

12.6. Functional Equations

In a number of important instances, R_N possesses a separability property. By this we mean that it takes the form

$$R_N = g(p, q_1) + g(p_1, q_2) + \cdots + g(p_{N-1}, q_N) + h(p_N). \qquad (12.6.1)$$

Intuitively this means that there is a return $g(p, q)$ at each stage of the process which depends only upon the current state p and the current decision q, plus a return for the terminal state of the system, p_N.

In place of using a direct application of calculus, or some other analytic technique, to maximize R_N over the choice of the q_i, we proceed in the following fashion. Introduce the function

$$f_N(p) = \max_{\{q_i\}} R_N, \tag{12.6.2}$$

defined for $N = 1, 2, \ldots$, and $p \in R$. In other words, we are trying to solve a whole class of problems at one time rather than one particular problem, a fundamental imbedding technique.

We see from the expression for R_1 that

$$f_1(p) = \max_{q_1}[g(p, q_1) + h(p_1)]. \tag{12.6.3}$$

In order to obtain a relation for f_N in terms of f_{N-1}, we argue as follows. Write

$$\begin{aligned} f_N(p) &= \max_{[q_1, q_2, \ldots, q_N]} R_N = \max_{q_1} \max_{[q_2, \ldots, q_N]} R_N \\ &= \max_{q_1}[g(p, q_1) + \max_{[q_2, \ldots, q_N]}\{g(p_1, q_2) + \cdots + h(p_N)\}]. \end{aligned} \tag{12.6.4}$$

By virtue of the definition of the function $f_N(p)$, we see that

$$\max_{[q_2, \ldots, q_N]}\{g(p_1, q_2) + \cdots + h(p_N)\} = f_{N-1}(p_1), \tag{12.6.5}$$

and hence that the expression on the right in (12.6.4) is equal to

$$\max_{q_1}[g(p, q_1) + f_{N-1}(p_1)].$$

Thus, using the expression for p_1, (12.6.4) yields

$$f_N(p) = \max_{q_1}[g(p, q_1) + f_{N-1}(T(p, q_1))], \qquad p \in S, \tag{12.6.6}$$

for $N \geqslant 2$, with $f_1(p)$ as in (12.6.3).

Exercises

1. Let $f_N(c) = \max_{x_i} \sum_{i=1}^{N} x_i^2$, where the x_i range over $\sum_{i=1}^{N} x_i = c$, $x_i \geqslant 0$. Show that

$$f_N(c) = \max_{0 \leqslant x_N \leqslant c} [x_N^2 + f_{N-1}(c - x_N)].$$

Use the fact that $f_N(c) = r_N c^2$ to determine r_N in terms of r_{N-1}.

2. Consider the maximization of $x_1 x_2 \cdots x_N$ over $\sum_{i=1}^{N} x_i = 1$, $x_i \geqslant 0$, and thus establish the arithmetic mean-geometric mean inequality, $(\sum_{i=1}^{N} y_i/N)^N \geqslant y_1 y_2 \cdots y_N$, $y_i \geqslant 0$, with equality only if the y_i are all equal.

12.7. Infinite Stage Process

Proceeding purely formally, let us consider the situation where the process is of unbounded duration. Write $f(p)$ as the formal limit of $f_N(p)$ as $N \to \infty$ and consider the functional equation

$$f(p) = \max_q [g(p, q) + f(T(p, q))]. \qquad (12.7.1)$$

We can now clearly simplify the notation by suppressing the subscript of q.

The usual problems now arise in connection with this functional equation, questions of existence and uniqueness of the solution of (12.7.1), and what is quite important, the relations between this solution and optimal choices in the original multistage decision process. Let us avoid any discussions of this nature here, interesting as they are. They occasionally involve some subtle reasoning and, in any case, are extraneous to the objectives of either this chapter or the book as a whole. In the miscellaneous exercises at the end of the chapter some methods for treating (12.7.1) are described. We wish to proceed to a discussion of (12.7.1) which illustrates the origin of quasilinearization. For our present purposes an intuitive understanding is all that is desired.

In attacking the equation of (12.7.1) along classical lines, we think immediately of employing that general factotum of analysis, the method of successive approximations. We can also, of course, utilize the Birkhoff–Kellogg method of fixed points in function space, a method which we have generally avoided despite its elegance because of its nonconstructive character. In general, in connection with equations associated with dynamic programming processes, uniqueness is a stickier question than existence.

Let $f_0(p)$ be some initial approximation and let the sequence $\{f_n(p)\}$, $n \geqslant 1$, be generated recursively by means of the relation

$$f_{n+1}(p) = \max_q [g(p, q) + f_n(T(p, q))], \qquad n = 0, 1, \ldots . \qquad (12.7.2)$$

Under various types of assumptions (see the exercises at the end of the chapter) that simultaneously guarantee existence and uniqueness of the solution of (12.7.1), we can establish the convergence of $f_n(p)$ to $f(p)$.

In many important cases this convergence, when it exists, will be monotone,

$$f_0 \leqslant f_1 \leqslant \cdots \leqslant f_n \leqslant f. \tag{12.7.3}$$

Consider, for example, the case where $g(p, q) \geqslant 0$ for $p \in R$, $q \in D$, and where $f_0(p) = \max_q g(p, q)$. The function f_n can then be conceived of as the return from an n-stage process of the type described in Sec. 12.4. Let us note that, as usual, extrapolation in n can be used to accelerate the convergence, or to provide a better initial approximation for a repetition of the approximation procedure. In many cases we think of the infinite stage process as providing an approximation to a finite stage process.

So far we have followed classical lines.

12.8. Policy

Let us now describe another approach to obtaining a solution of (12.7.1) which has no counterpart in classical analysis. For reasons that will become clear, we call it "approximation in policy space." Observe to begin with in actuality *two* functions are determined by (12.7.1), the function $f(p)$ which we call the "return function" and the function $q(p)$ which we call the "policy function." This latter function, which may be multivalued, is determined by the value, or values, of q which yield the maximum in (12.7.1).

The function $q(p)$ is called a "policy" because it determines the choice of a decision in terms of the current state of the system. This agrees with the usual intuitive meaning of the term. The policy which maximizes the total return is called the optimal policy. Again, it need not be unique.

Once $f(p)$ has been found, $q(p)$ is obtained by maximizing the expression $[g(p, q) + f(T(p, q))]$. With $q(p)$ obtained, $f(p)$ is determined by straightforward iteration,

$$f(p) = g(p, q) + g(T(p, q), q(T(p, q))) + \cdots, \tag{12.8.1}$$

which is to say by carrying out the decision process.

Consequently, we can focus either on $f(p)$ or $q(p)$ as far as deriving the solution of (12.7.1) is concerned. Let us then examine the possibility of approximating to the optimal policy, rather than to the return function. In many cases, we possess an intuitive hold on the policy function from a knowledge of the underlying decision process. In general, as might be imagined, an interplay of the two types of approximation will provide a powerful leverage.

12.9. Approximation in Policy Space

Let $q_0(p)$ be an initial guess for the optimal policy and let $f_0(p)$ be determined by the equation

$$\begin{aligned} f_0(p) &= g(p, q_0) + f_0(T(p, q_0)) \\ &= g(p, q_0) + g(T(p, q_0), q_0(T(p, q_0))) + \cdots. \end{aligned} \tag{12.9.1}$$

Observe that $f_0(p)$ is necessarily a return function since it corresponds to the use of a particular policy. When the method of successive approximations is used in a routine fashion, it is not necessarily the case that the functions obtained are return functions. We shall assume that we have imposed suitable conditions on the functions g and T so that the equation has a unique solution obtained by direct iteration; see the exercises following.

To obtain a second approximation, let $q_1(p)$ be determined by the condition that it maximize the expression

$$g(p, q) + f_0(T(p, q)), \tag{12.9.2}$$

and let a new return function $f_1(p)$ be determined using this policy. Then f_1 satisfies the equation

$$f_1(p) = g(p, q_1) + f_1(T(p, q_1)), \tag{12.9.3}$$

Intuitively, we are trying to determine the best initial decision, knowing that we will be using the old policy from the second stage on. We then use this best initial decision as a policy for the entire process. Let us now examine the relations between f_0 and f_1.

We have, by virtue of the method used to determine $q_1(p)$,

$$f_0(p) = g(p, q_0) + f_0(T(p, q_0) \leqslant g(p, q_1) + f_0(T(p, q_1)) \tag{12.9.4}$$

for all $p \in R$. Hence, we suspect that

$$f_0(p) \leqslant f_1(p), \tag{12.9.5}$$

since f_1 is determined by the equality of (12.9.3).

We see then that monotonicity of convergence depends entirely upon the nature of the solution of (12.9.3), f_1, as a function of g. In many cases of importance, the iterative solution of (12.9.1) shows that f_1 has the proper monotonicity property. In other cases, more sophisticated argumentation is required; see Chapter 10.

Exercises

1. Consider the equation of (12.9.1) where we assume that $T(p, q)$ is a contracting transformation, $|T(p, q)| \leqslant a|p|$ for all $p \in R$ and all q, $0 < a < 1$, and that $|g(p, q)| \leqslant h(p)$ where $\sum_n h(a^n p) < \infty$. Show that (12.9.1) has a unique solution given by direct iteration.
2. Establish a corresponding result for $f(p) = g(p, q) + h(p) f(T(p, q))$ under the assumption that $|h(p)| < b < 1$ for all $p \in R$ under various assumptions concerning T.
3. Study the monotonicity of the solution of (12.9.1) making various positivity and monotonicity assumptions concerning g and T.

12.10. Discussion

With the foregoing in mind, we can understand why we were so eager to convert an equation of the familiar form

$$u' = u^2 + a(t), \qquad u(0) = c, \tag{12.10.1}$$

into a rather strange-looking equation involving a maximum operation and then to introduce what corresponds to a policy. Furthermore, we can now see the reason for the strong interest previously expressed in monotone operators and differential inequalities.

12.11. Calculus of Variations as a Multistage Decision Process

So far we have considered that decisions were made at discrete times, a realistic assumption. It is, however, also of considerable interest to examine the fiction of continuous decision processes. In particular, we wish to demonstrate that we can profitably regard the calculus of variations as an example of a multistage decision process of continuous type. In consequence of this, dynamic programming provides a number of new conceptual, analytic, and computational approaches to classical and modern variational problems, particularly to those arising in control processes.

To illustrate the basic idea, which is both quite simple and natural from the standpoint of a control process, let us consider the scalar functional

$$J(u) = \int_0^T g(u, u')\, dt. \tag{12.11.1}$$

The problem of minimizing $J(u)$ over all u satisfying the initial condition $u(0) = c$ leads along classical lines as we know to the task of solving the Euler equation

$$\frac{\partial g}{\partial u} - \frac{d}{dt}\left(\frac{\partial g}{\partial u'}\right) = 0, \tag{12.11.2}$$

subject to the two-point condition

$$u(0) = c, \qquad \left.\frac{\partial g}{\partial u'}\right|_{t=T} = 0. \tag{12.11.3}$$

This equation, as we have already seen, is a variational equation obtained by considering the behavior of the functional $J(u + w)$ for "all" small w where u is the desired minimizing function. This procedure of examining the neighborhood of the extremal in function space is a natural generalization of that used in calculus in the finite-dimensional case. As in the finite-dimensional case, there can be considerable difficulty first in solving the variational equation and then in distinguishing the absolute minimum from other stationary points.

Let us now pursue an entirely different approach motivated by the theory of dynamic programming. In particular, it is suggested by the applications of dynamic programming to the study of deterministic control processes. In place of thinking of a curve $u(t)$ as a locus of points, let us take it to be an envelope of tangents; see Figs. 12.2 and 12.3. Ordinarily, we determine a point on the curve by the coordinates

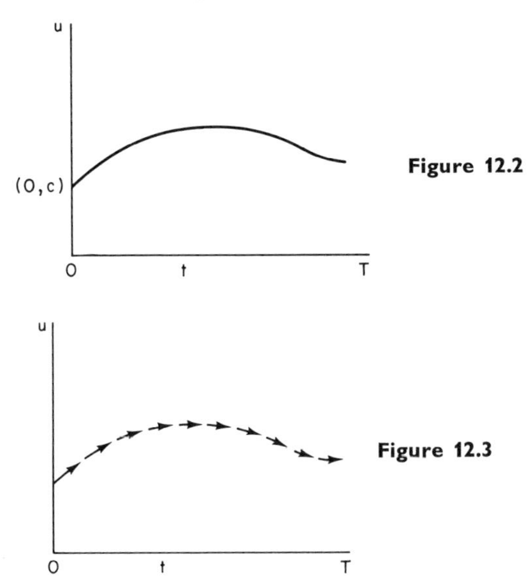

Figure 12.2

Figure 12.3

$(t, u(t))$. However, we can equally well trace out the curve by providing a rule for determining the slope u' at each point (t, u) along the path. The determination of the minimizing curve $u(t)$ can thus be regarded as a multistage decision process in which it is necessary to choose a direction at each point along the path. Motivation of this approach in the domain of pursuit processes is easily seen, or equivalently, in the determination of geodesics, in the analysis of multistage investment processes, or in the study of optimal growth processes in mathematical economics.

12.12. A New Formalism

Let us use the foregoing concept of a minimization process to obtain a new analytic approach to variational problems. See Fig. 12.4. Let us

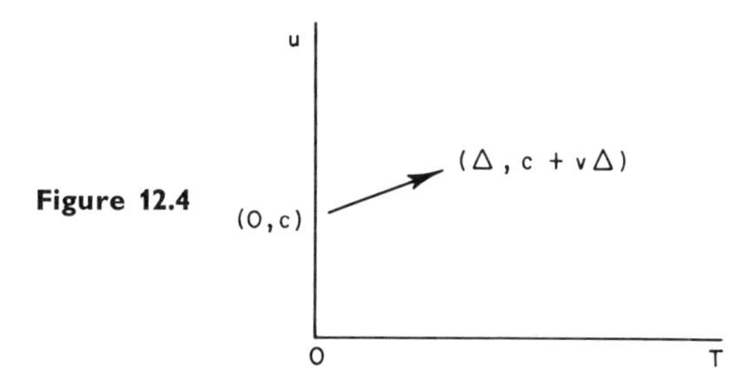

Figure 12.4

denote the minimum value of $J(u)$ as defined by (12.11.1) (assumed to exist) by $f(c, T)$. Thus, we introduce the function

$$f(c, T) = \min_u J(u), \tag{12.12.1}$$

defined for $T \geqslant 0$ and $-\infty < c < \infty$. Since we are now interested in policies, the initial state must be considered to be a variable.

For the problem (12.11.1), a suitable initial condition is $f(c, 0) = 0$. The situation is a bit more complicated when there is a two-point boundary condition; see Exercise 4.

Let us first proceed formally (as in the derivation of the Euler equation) supposing that all partial derivatives exist, that all limiting procedures are valid, etc.

Let $v = v(c, T)$, as in the foregoing figure, denote the initial direction; $v = u'(0)$, which clearly depends on c and T. This is the missing initial value in (12.11.2), an essential quantity. Writing

$$\int_0^T = \int_0^\varDelta + \int_\varDelta^T, \tag{12.12.2}$$

a convenient separability property of the integral, we see that for any initial $v(c, T)$ we have

$$f(c, T) = \Delta g(c, v) + O(\Delta^2) + \int_{\Delta}^{T} . \qquad (12.12.3)$$

Here Δ is an infinitesimal. We now argue as follows. Regardless of how $v(c, T)$ was chosen, we are going to proceed from the new point $(\Delta, c + v\Delta)$ so as to minimize the remaining integral $\int_{\Delta}^{T} g(u, u')\, dt$. But

$$\min_{u} \int_{\Delta}^{T} g(u, u')\, dt = f(c + v\Delta, T - \Delta) \qquad (12.12.4)$$

by definition of the function f. Hence, for any initial choice of v, we have

$$f(c, T) = g(c, v)\Delta + f(c + v\Delta, T - \Delta) + O(\Delta^2). \qquad (12.12.5)$$

This is an example of the "principle of optimality" for multistage decision processes, quoted below in Sec. 12.13.

It remains to choose $v(c, T)$ appropriately. Clearly, v should be chosen to minimize the right-hand side of (12.12.5). Thus, we obtain the equation

$$f(c, T) = \min_{v}[g(c, v)\Delta + f(c + v\Delta, T - \Delta)] + O(\Delta^2). \qquad (12.12.6)$$

Expanding the appropriate terms above in powers of Δ and letting $\Delta \to 0$, we obtain the partial differential equation

$$f_T = \min_{v}[g(c, v) + vf_c]. \qquad (12.12.7)$$

We have noted above that $f(c, 0) = 0$, an initial condition. Thus we have transformed the original variational problem of minimizing $J(u)$ in (12.11.1) into that of solving a nonlinear partial differential equation subject to an initial condition.

The foregoing is a cavalier approach in the spirit of the usual first derivation of the Euler equation. We leave it to the reader to spot all the irregularities.

Exercises

1. Show that the problem of minimizing

$$J(u) = \int_{0}^{T} g(u, v)\, dt,$$

where u and v are connected by the differential equation $u' = h(u, v)$, $u(0) = c$, leads to the equation

$$f_T = \min_v [g(c, v) + h(c, v) f_c],$$

$f(c, 0) = 0$.

2. Obtain the corresponding equation for

$$f(c, T) = \min_x \int_0^T g(x, x')\, dt,$$

where x is an N-dimensional vector, and $x(0) = c$.

3. Obtain the corresponding equation for

$$f(c, T) = \min_y \int_0^T g(x, y)\, dt,$$

where $x' = h(x, y)$, $x(0) = c$.

4. What is a correct initial condition for $f(c, T)$ when we set

$$f(c, T) = \min_u \int_0^T g(u, u')\, dt$$

subject to $u(0) = c$, $u(T) = c_1$?

5. Obtain an equation for

$$f(a, c) = \min \int_a^T u'^2\, dt,$$

where

$$\int_a^T u^2\, dt = 1,$$

$u(a) = c$. What is a correct initial condition?

12.13. The Principle of Optimality

The reasoning employed in the foregoing section to derive (12.12.5) is a particular case of the following general principle:

Principle of Optimality. *An optimal policy has the property that whatever the initial state and initial decision are, the remaining decisions must constitute an optimal policy with regard to the state resulting from the first decision.*

A proof by contradiction is immediate in most cases. Difficulties in the application of this principle reside in the demonstration that an optimal policy exists and in the selection of appropriate state variables.

12.14. Quadratic Case

A case of particular importance, for reasons we have already discussed in Chapter 8, Volume I, is that where the integrand is quadratic. Consider first the scalar case where we wish to minimize the functional

$$J(u) = \int_0^T (u'^2 + b(t)\, u^2)\, dt, \tag{12.14.1}$$

subject to the condition $u(0) = c$. Rigorous aspects of what follows are discussed in Section 12.23. To take care in a simple fashion of the dependence of the integrand on t, we count time backwards and write

$$f(c, a) = \min_u \int_a^T (u'^2 + b(t)\, u^2)\, dt. \tag{12.14.2}$$

Here $0 \leqslant a \leqslant T$, and the initial condition is $u(a) = c$. The preceding formalism yields the equation

$$-\frac{\partial f}{\partial a} = \min_v \left[v^2 + b(a)\, c^2 + v \frac{\partial f}{\partial c}\right], \tag{12.14.3}$$

with the initial condition $f(c, T) = 0$. The minimum with respect to v is easily obtained, leading to the equation

$$-\frac{\partial f}{\partial a} = b(a)\, c^2 - \frac{1}{4}\left(\frac{\partial f}{\partial c}\right)^2, \qquad f(c, T) = 0. \tag{12.14.4}$$

Nonlinear partial differential equations have their obvious disadvantages. Hence, it is particularly fortunate, but not unexpected, that we can reduce the problem of solving (12.14.4) to that of solving an ordinary differential equation of Riccati type.

We observe, either directly by a change of dependent variable, $u = vc$, or by reference to the associated Euler equation and the results of Chapter 8, Volume I, that $f(c, a)$ is proportional to c^2,

$$f(c, a) = r(a)\, c^2, \tag{12.14.5}$$

where $r(a)$ is a function only of a.

Using this representation, (12.14.4) reduces to the ordinary differential equation

$$-r'(a) = b(a) - r(a)^2, \qquad r(T) = 0, \tag{12.14.6}$$

a simple initial-value problem.

Exercises

1. Obtain the corresponding equations for the case where

$$J(u) = \int_a^T [u'^2 + b_1(t)u^2 + 2b_2(t)u]\, dt.$$

2. Obtain the corresponding equation and initial value when $J(u)$ is as in (12.11.1) and u is subject to a two-point condition $u(a) = c$, $u(T) = c_1$.

3. Obtain the solution of the Euler equation from a knowledge of $r(a)$.

4. Show directly from the discussion of Chapter 8 that $r(a)$ satisfies (12.14.6).

5. Obtain the optimal policy in terms of $r(a)$.

6. Let

$$f(c, T) = \min_u \int_0^T (u'^2 + u^2)\, dt, \qquad u(0) = c.$$

What are the asymptotic behaviors of $f(c, T)$ and $v(c, T)$ as $T \to \infty$?

7. Can one obtain the function

$$f(c) = \min_u \int_0^\infty (u'^2 + u^2)\, dt, \qquad u(0) = c,$$

as the limit of $f(c, T)$ as $T \to \infty$?

8. Show that the solution of the Riccati equation exists and is uniformly bounded for all $0 \leqslant a \leqslant T < \infty$.

12.15. Multidimensional Case

Let us now carry through the analytic details for the multidimensional case. Let

$$J(x) = \int_a^T [(x', x') + (x, B(t)x)]\, dt, \tag{12.15.1}$$

with $x(a) = c$. Set

$$f(c, a) = \min_x J(x). \tag{12.15.2}$$

Then, as before, we obtain the equation

$$-\frac{\partial f}{\partial a} = \min_v [(v, v) + (c, B(a)c) + (\text{grad}\, f, v)], \tag{12.15.3}$$

where grad f is, as usual, the vector whose components are $\partial f/\partial c_i$, with c_i the ith component of c. The minimizing v is given by

$$v = -(\text{grad}\, f)/2. \tag{12.15.4}$$

Using this expression for v, (12.15.3) reduces to the nonlinear equation

$$-\frac{\partial f}{\partial a} = (c, B(a)c) - (\text{grad}\, f, \text{grad}\, f)/4, \tag{12.15.5}$$

subject to $f(c, T) = 0$. We now employ the fact that

$$f(c, a) = (c, R(a)c), \tag{12.15.6}$$

a quadratic form in the components of c. This can be seen from the associated Euler equation, or by analogy with the scalar case of Section 12.14. We see that $R(a)$ satisfies the matrix Riccati differential equation

$$-R'(a) = B(a) - R(a)^2, \qquad R(T) = 0, \tag{12.15.7}$$

again an initial-value problem.

Exercises

1. Consider the case where

$$J(x) = \int_0^T [(x', x') + (x, B(t)x) + 2(g(t), x)]\, dt.$$

2. Determine

$$\lim_{T\to\infty} \min_x \left(\int_0^T (x', x') + (x, Ax) \right) dt, \qquad x(0) = c,$$

where A is a positive definite matrix.

3. How does one obtain the solution of the original variational problem in (12.15.1) in terms of $R(a)$?

4. Show that $R(a)$, the solution of (12.15.7), exists and is uniformly bounded for all $0 \leqslant a \leqslant T < \infty$.

12.16. Computational Feasibility

If x is N-dimensional, $R(a)$ is an $N \times N$ matrix. Since it is symmetric, we see that (12.12.7) constitutes a set of $N(N+1)/2$ simultaneous ordinary differential equations for the elements of $R(a)$, subject to initial conditions. Analytically, this resolves the problem. How feasible, however, is this approach computationally?

If $N = 10$, there are 55 such equations; $N = 25$ leads to 325 equations; $N = 100$ leads to 5050. None of these cases represent exorbitant demands on a contemporary computer.* This increase in capacity over computers of ten years ago is quite important since it also enables us to handle many classes of partial differential equations in the same direct fashion, using various simple approximation methods. We shall return to this point in Chapter 15.

Let us compare the approach using the Riccati equation with the approach described in the previous pages. The Euler equation, obtained using the calculus of variations, is

$$x'' - B(t)x = 0, \qquad x(a) = c, \quad x'(T) = 0. \tag{12.16.1}$$

The usual approach to the solution of this equation leads to the task of solving a set of N simultaneous linear algebraic equations. This is always a touchy affair as we have repeatedly emphasized. We will enlarge upon this later in the chapter. If this operation has to be done repeatedly in connection with the use of successive approximations of the type described in connection with the application of quasilinearization to more general functionals, there is a considerable risk of numerical error overwhelming the procedure, or in any case of blunting the precision of quasilinearization.

12.17. Stability

Another important point to mention is that the procedure of (12.16.1), derived from the calculus of variations, can be numerically unstable if routine methods are employed while that associated with the Riccati equation, the algorithm obtained from dynamic programming is stable using the ordinary computational approach. This does not mean that

* ca. 1972.

the classical procedure cannot be used, but it does mean that a certain amount of care is required at all times.

To illustrate this, consider perhaps the simplest variational problem,

$$\min_u \int_0^T (u'^2 + u^2)\, dt, \tag{12.17.1}$$

with $u(0) = 1$. The Euler equation is

$$u'' - u = 0, \qquad u(0) = 1, \quad u'(T) = 0 \tag{12.17.2}$$

with the explicit solution

$$u = \frac{\cosh(t - T)}{\cosh T}. \tag{12.17.3}$$

The associated initial condition is

$$u'(0) = -\tanh T. \tag{12.17.4}$$

The desired solution has the form in Fig. 12.5. Suppose, however, that we try to integrate the equation

$$u'' - u = 0, \qquad u(0) = 1, \quad u'(0) = -\tanh T, \tag{12.17.5}$$

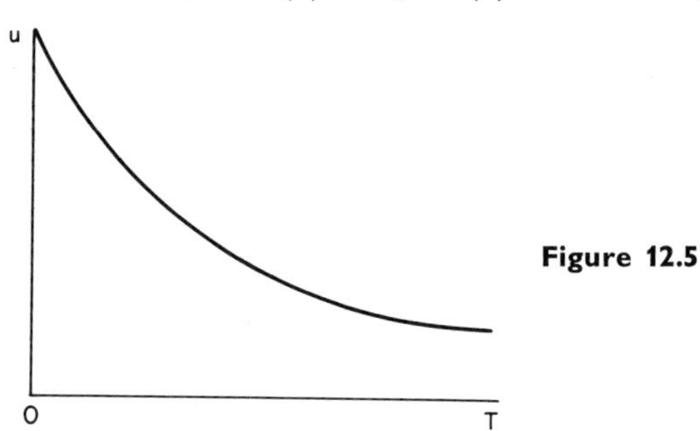

Figure 12.5

numerically using a standard algorithm for the digital computer. As we have already discussed, this operation can be considered to be equivalent to solving a linear homogeneous equation

$$u'' - u = e(t), \tag{12.17.6}$$

where $e(t)$ is a forcing term corresponding to the effects of discretization, round-off error, and so forth. The effect of these small errors is essentially to introduce an extraneous term of the form ϵe^t in (12.17.4). Hence, the calculated solution has the form in Fig. 12.6. For small T, the effect is

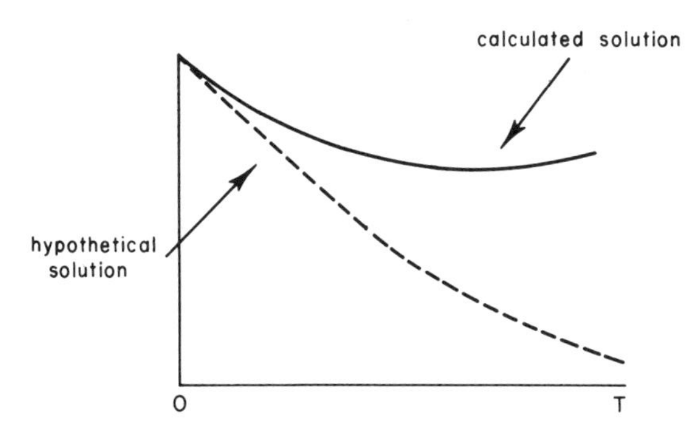

Figure 12.6

negligible. For large T, it becomes overwhelming. In multidimensional cases, the effects can be considerably magnified, particularly when one characteristic root is significantly different in magnitude from the others.

Let us turn to the dynamic programming approach. Writing

$$r(T) = \min_u \int_0^T (u'^2 + u^2)\, dt, \qquad u(0) = 1, \tag{12.17.7}$$

we know that $r(T)$ satisfies the Riccati equation

$$r'(T) = 1 - r(T)^2, \qquad r(0) = 0. \tag{12.17.8}$$

As we have pointed out in Sec. 4.21, Exercises, in Volume I, this is a stable equation under perturbations in either the structure or the initial condition. The same holds for the matrix Riccati equation

$$R'(T) = A - R^2(T), \qquad R(0) = 0. \tag{12.17.9}$$

This stability is to be expected in view of the self-correcting nature of the associated continuous control, or equivalently multistage decision, process.

12.18. Computational Feasibility: General Case—I

Let us now briefly examine the application of the foregoing techniques to the numerical solution of the problem of minimizing the more general functional

$$J(u) = \int_0^T g(u, u')\, dt, \tag{12.18.1}$$

subject to the initial condition $u(0) = c$. The dynamic programming approach yields the nonlinear partial differential equation

$$f_T = \min_v [g(c, v) + v f_c], \qquad f(c, 0) = 0, \tag{12.18.2}$$

for $f(c, T) = \min_u J(u)$. Assuming that $g(c, v)$ is convex in v for all c, the unique minimizing value is determined by the equation

$$g_v + f_c = 0. \tag{12.18.3}$$

Solving for v, we can write (12.18.2) in the form

$$f_T = h(c, f_c), \qquad f(c, 0) = 0. \tag{12.18.4}$$

Standard techniques involving the use of characteristics may be employed to treat this equation. Alternatively, if the policy alone is what is desired, we may eliminate f between (12.18.2) and (12.18.3). With v determined by (12.18.3), (12.18.2) reads

$$f_T = g(c, v) + v f_c = g(c, v) - v g_v \,. \tag{12.18.5}$$

Differentiating this last equation with respect to c, we obtain

$$f_{Tc} = g_c + g_v v_c - v_c g_v - v g_{cv} - v g_{vv} v_c = g_c - v g_{cv} - v g_{vv} v_c \,. \tag{12.18.6}$$

Differentiating (12.18.3) with respect to T, we have

$$f_{cT} = -g_{vv} v_T \,. \tag{12.18.7}$$

Equating f_{cT} and f_{Tc}, we derive the equation

$$v_T = v v_c + (v g_{cv} - g_c)/g_{vv} \,. \tag{12.18.8}$$

The value of $v(c, T)$ for $T = 0$ is determined by the function of c which minimizes $g(c, v)$.

Once again the theory of characteristics may be employed, to continue from here both analytically and computationally.

Exercise

1. Using the theory of characteristics, or otherwise, derive the Euler equation from (12.18.2) or (12.18.3).

12.19. Computational Feasibility: General Case—II

In place of the foregoing approach, we can employ a method which is based upon use of a discrete version of the continuous variational problem. Instead of one of the conventional difference schemes to handle (12.18.2) let us write

$$f(c, T + \Delta) = \min_{v}[g(c, v)\Delta + f(c + v\Delta, T)], \qquad f(c, 0) = 0, \qquad (12.19.1)$$

where T now assumes only the values 0, Δ, 2Δ,..., and c assumes values in some interval $[c_1, c_2]$. The function $f(c, T)$ is stored in this interval for each value of T either by means of a set of values at grid points or by means of an interpolation formula. The minimization is determined by a direct search over a discrete set of values of v.

12.20. The Curse of Dimensionality

There are a number of questions connected with this approach which are discussed in detail in the references cited at the end of the chapter. The procedure is easily followed in a routine fashion when the dimension of c is low, say 1, 2, or 3. Difficult and interesting storage and retrieval questions arise as the dimension increases, say 4, 5, 6. As the dimension increases still further, various techniques of successive approximations must be employed.

The significant feature of this is that successive approximations is now employed not so much to yield linearity as to furnish decompositions which substantially lower the dimension of the component processes.

12.21. Constraints

In a number of important processes occurring in economics and engineering, free variations are not permitted. Thus, for example, we may be confronted with the problem of minimizing

$$J(u) = \int_0^T g(u, u')\, dt \qquad (12.21.1)$$

subject to the constraint $|u'| \leqslant 1$, $0 \leqslant t \leqslant T$, with $u(0) = c$.

Proceeding formally, we obtain the nonlinear equation

$$f_T = \min_{|v| \leqslant 1} [g(c, v) + vf_c], \qquad f(c, 0) = 0, \qquad (12.21.2)$$

for the function $f(c, T) = \min J(u)$. A point well worth mentioning is that the associated computational algorithm

$$f(c, T + \Delta) = \min_{|v| \leqslant 1} [g(c, v)\Delta + f(c + v\Delta, T)] \tag{12.21.3}$$

works even more effectively when the constraint is present since the search for the minimum over v is substantially reduced.

The task of dealing in a rigorous fashion with constrained variational problems, using either the calculus of variations or dynamic programming, is, however, far more difficult than treating the unconstrained case. References will be found at the end of the chapter.

In particular, in treating (12.21.2) we often encounter some interesting shock-like phenomena. The function f_c can be discontinuous along certain curves in the (c, T) plane, an analog of a shock in hydrodynamics.

12.22. Two-point Boundary Value Problems

We have previously discussed an analytic and computational approach to the two-point boundary value problem

$$u'' = k(u, u'), \qquad u(0) = c, \quad u(T) = c_1, \tag{12.22.1}$$

using quasilinearization.

If the equation in (12.22.1) is the Euler equation connected with a variational problem

$$\min_u \int_0^T g(u, u')\, dt, \qquad u(0) = c, \quad u(T) = c_1, \tag{12.22.2}$$

we can employ an algorithm based on dynamic programming instead of a procedure based on successive approximations. An interesting point is that we can apply dynamic programming to a discrete version to obtain an excellent initial approximation, and then quasilinearization to obtain a highly accurate result if we wish.

In the following chapter we present a method based upon invariant imbedding which is more powerful in the sense that it does not require that (12.22.1) be derived from a variational problem.

12.23. Rigorous Aspects

As in the case of the calculus of variations, it is essential to examine the validity of the equations obtained by formal analytic manipulations.

There are several ways in which we can approach the problem of providing a rigorous basis for the foregoing results. In the quadratic case, the path is easy. Using the results of Chapter 8, Volume I, we can evaluate the quantity

$$(c, R(a)c) = \min_x \left[\int_a^T [(x', x') + (x, B(t)x)] \, dt \right], \qquad (12.23.1)$$

where $x(0) = c$, explicitly in terms of the solutions of $x'' - B(t)x = 0$, and establish by a simple direct calculation that $R(a)$ satisfies the matrix Riccati equation of Sec. 12.15.

The more general problem of minimizing

$$J(x) = \int_0^T g(x, x') \, dt, \qquad (12.23.2)$$

where $x(0) = c$, and x is an N-dimensional vector, may be handled in an analogous fashion using classical tools of the calculus of variations. Under suitable assumptions concerning g, a solution of the Euler equation can be shown to exist and be unique. Furthermore, it can be established that $\min_x J(x) = f(c, T)$ is a sufficiently differentiable function of c and T. This is particularly simply accomplished for T small. A direct calculation then shows that $f(c, T)$ satisfies the desired nonlinear partial differential equation.

A third approach which is interesting from the point of view of both analysis and numerical calculation is the following. Consider, to simplify, the problem of minimizing the scalar functional

$$J(u, v) = \int_0^T g(u, v) \, dt, \qquad (12.23.3)$$

subject to

$$\frac{du}{dt} = h(u, v), \qquad u(0) = c. \qquad (12.23.4)$$

To this continuous proccss, we associate a discrete variational problem: minimize with respect to the v_n the function of N variables,

$$J(u_n, v_n) = \sum_{n=0}^{N-1} g(u_n, v_n)\Delta, \qquad (12.23.5)$$

where the u_n and v_n are connected by the relation

$$u_{n+1} = u_n + h(u_n, v_n)\Delta, \qquad u_0 = c, \qquad (12.23.6)$$

$n = 0, 1, \ldots, N - 1$.

Here $u_n \cong u(n\Delta)$, $v_n \cong v(n\Delta)$. If we define the function

$$f_n(c) = \min_{\{v_n\}} J(u_n, v_n), \tag{12.23.7}$$

$n = 0, 1, \ldots, -\infty < c < \infty$, we readily obtain the functional equation

$$f_n(c) = \min_v [g(c, v)\Delta + f_{n-1}(c + h(c, v)\Delta)], \tag{12.23.8}$$

for $n \geqslant 1$, with

$$f_0(c) = \min_v [g(c, v)\Delta]. \tag{12.23.9}$$

If we set $N\Delta = T$, it is plausible that

$$\lim_{N\to\infty} f_N(c) = f(c, T), \tag{12.23.10}$$

and indeed this limiting behavior can be established under various reasonable assumptions.

This result is of analytic interest, and also essential if the computational method of Sec. 12.19 is to be used with confidence.

12.24. The Equation $u_t = \varphi(x, u_x)$

Let us now turn our attention to the application of quasilinearization to the equation

$$u_t = \varphi(x, u_x), \qquad u(x, 0) = h(x). \tag{12.24.1}$$

If $\varphi(x, u_x)$ is uniformly convex in u_x, we can write

$$\begin{aligned} \varphi(x, u_x) &= \max_w [\varphi(x, w) + (u_x - w)\, \varphi_w(x, w)] \\ &= \max_w [\varphi - w\varphi_w + u_x\varphi_w]. \end{aligned} \tag{12.24.2}$$

Hence, (12.24.1) may be written

$$u_t = \max_w [(\varphi - w\varphi_w) + u_x\varphi_w]. \tag{12.24.3}$$

The function $f(c, T)$ defined by

$$f(c, T) = \min \left[\int_0^T g(z, z')\, dt + h(z(T))\right], \tag{12.24.4}$$

where $z(0) = c$, satisfies the equation of similar structure,

$$f_T = \min_v [g(c, v) + vf_c], \qquad f(c, 0) = h(c). \tag{12.24.5}$$

If we identify x with c, and make the changes of variable

$$\varphi_w = v, \qquad \varphi - w\varphi_w = g(c, v), \tag{12.24.6}$$

we see that we can express the solution of (12.24.1) in the form of the right-hand side of (12.24.4). From this explicit expression, many properties of the solution of (12.24.1) can be deduced.

Furthermore, the monotonicity of the operator $L(g)$ defined by

$$f_T - vf_c = g \tag{12.24.7}$$

enables us to deduce accurate upper bounds by means of approximation in policy space. Using variational problems dual to that in (12.24.4) (see Chapter 9), we can obtain corresponding partial differential equations from which we can derive lower bounds for the solution of (12.24.1).

Exercises

1. Obtain corresponding results for $u_t = \varphi(x, t, u_x)$.
2. Carry through the representation process for $u_t = u_x^2$. From this, deduce a representation for the solution of $w_t = ww_x$.

12.25. The Equation $u_t = \varphi(x, u_{xx})$

We began the chapter by considering a discrete multistage decision process and then progressed to the continuous multistage decision process in order to make contact with the calculus of variations and nonlinear partial differential equations of the form

$$f_T = \min_v [g(c, v) + (\operatorname{grad} f, v)]. \tag{12.25.1}$$

If we introduce the concept of a multistage decision process of stochastic type, we obtain a wide-ranging extension of the previous theory. In particular, if we consider the appropriate continuous stochastic control process, we can introduce nonlinear equations of the form

$$f_T = \min_v [g(c, v) + h(c, v) f_c] + f_{cc} \tag{12.25.2}$$

together with their multidimensional versions.

In this way we can apply the concepts of dynamic programming, and particularly approximation in policy space, to the study of nonlinear partial differential equations of parabolic and elliptic type. Reference to analyses of this nature will be found at the end of the chapter.

In the following chapter, we will show how the theory of invariant imbedding can be applied in an analogous fashion to nonlinear partial differential equations of hyperbolic type.

12.26. Generalized Semigroup Theory and Nonlinear Equations

A large, and very important, part of the corpus of the classical theory of ordinary and partial differential equations can be considered to be part of the theory of semigroups of operations. In particular, we can begin with an operator equation of the form

$$u_t = Au + g, \qquad u(0) = h, \tag{12.26.1}$$

and subsume many particular results under a study of common properties of the solution of (12.26.1).

Thus, the equation may be an ordinary differential equation of the form

$$\frac{dx}{dt} = Ax + g, \qquad x(0) = c. \tag{12.26.2}$$

where A is a matrix, or, it may assume the form

$$u_t = u_{xx} + g, \qquad u(x, 0) = h(x), \tag{12.26.3}$$

and so on.

The question arises as to how to construct a meaningful extension of this linear theory which can be used to handle classes of nonlinear equations. One answer, suggested by the foregoing, is to consider the quasilinear equation

$$u_t = \max_v [A(v)u + g(v)], \qquad u(0) = h, \tag{12.26.4}$$

associated with a continuous dynamic programming process as a prototype equation. One bonus of this approach is that the calculus of variations is included automatically.

A natural extension of semigroups of transformations is thus seen to be semigroups of decisions. Introducing stochastic and adaptive features, we can greatly extend the classes of equations included in this theory.

12.27. Ill-conditioned Systems

In Chapter 2, Volume I, we pointed out that the task of obtaining the numerical solution of

$$Ax = b \tag{12.27.1}$$

can be exceedingly difficult, and indeed essentially impossible, if A is sufficiently ill-conditioned. We can turn the difficulty if we add to (12.27.1) some further information concerning the nature of x. Following Tychonov, one way to do this is to consider the new problem of minimizing the expression

$$g(x) = (Ax - b, Ax - b) + \varphi(x), \tag{12.27.2}$$

where $\phi(x)$ is a suitably chosen function. There are several ways of choosing $\varphi(x)$. If $\varphi(x) = 0$, we have the original problem. In some cases, as in numerical inversion of the Laplace transform, see Chapter 16, we have an approximate idea of the solution, a vector c. We can then take

$$\varphi(x) = \lambda(x - c, x - c), \tag{12.27.3}$$

where $\lambda > 0$ must be carefully selected.

In other cases, we use a self-consistent approach. We often know that the components of x, $x_1, x_2, \ldots, x_N$, are smooth functions of the index. Hence, we take

$$\varphi(x) = \lambda[(x_2 - x_1)^2 + (x_3 - x_2)^2 + \cdots + (x_N - x_{N-1})^2]. \tag{12.27.4}$$

Once the problem has been formulated in either of these fashions, we can apply dynamic programming techniques by regarding the choice of the minimizing x as a multistage decision process in which we choose first x_N, then x_{N-1}, and so on.

12.28. Adaptive Polygonal Approximation

Let us consider the problem of obtaining a polygonal approximation to a function $u(t)$ defined over an interval $[0, T]$. (See Fig. 12.7.) Observe that initially we make no demand that the polygonal approximation be connected at the transition points, t_1 and t_2. If we fix the points $t_1, t_2, \ldots$, the problem of finding the best mean-square approximation is routine. Let us consider the more interesting and important case where the t_i are to be chosen in some expeditious fashion. Since their location will depend generally on the nature of the function u, we see the origin of the name "adaptive."

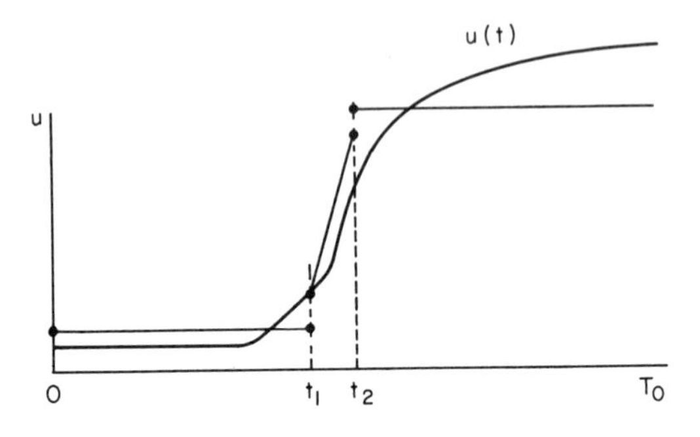

Figure 12.7

Let us then consider the general problem where there are N points of subdivision. We wish then to minimize the expression

$$\sum_{i=0}^{N} \left[\int_{t_i}^{t_{i+1}} (u(t) - a_i - b_i t)^2 \, dt \right], \tag{12.28.1}$$

with respect to the parameters a_i, b_i, and t_i. Here $t_0 = 0$, $t_{N+1} = T_0$. Observe that we can conceive of this minimization as a multistage decision process in which first t_N is chosen, then the best mean-square approximation to $u(t)$ in $[t_N, T_0]$ is determined, then t_{N-1} is chosen, and so on. Let us then introduce the function

$$f_N(T) = \min_{[a_i, b_i, t_i]} \left[\sum_{i=0}^{N} [\cdots] \right], \tag{12.28.2}$$

defined for $N = 0, 1, \ldots, 0 \leqslant T \leqslant T_0$. We have

$$f_0(T) = \min_{[a_1, b_1]} \int_0^T (u(t) - a_1 - b_1 t)^2 \, dt, \tag{12.28.3}$$

a function readily obtained. Generally, define

$$\Delta(r, s) = \min_{a,b} \left[\int_r^s (u(t) - a - bt)^2 \, dt \right] \tag{12.28.4}$$

for $0 \leqslant r \leqslant s \leqslant T_0$.

Then, viewing the minimization as a multistage process, we readily obtain the relation

$$f_N(T) = \min_{0 \leqslant t_N \leqslant T} [\Delta(t_N, T) + f_{N-1}(t_N)], \tag{12.28.5}$$

for $N = 1, 2, \ldots$. This is a simple algorithm for the digital computer.

Exercises

1. Calculate $f_N(T)$ in explicit analytic form for $u(t) = t^k$, $k \geqslant 1$.

2. What is the asymptotic behavior of $f_N(T)$ as $N \to \infty$, and as $T \to \infty$, in the case where $u = t^k$?

3. Obtain the corresponding functional equations for the case where we wish to minimize

$$\sum_{i=0}^{N} \left[\int_{t_i}^{t_{i+1}} | u(t) - a_i - b_i t | \, dt \right],$$

$$\sum_{i=0}^{N} \max_{t_i \leqslant t \leqslant t_{i+1}} | u(t) - a_i - b_i t |,$$

$$\max_i \int_{t_i}^{t_{i+1}} [u(t) - a_i - b_i t]^2 \, dt.$$

4. Modify the foregoing procedure to consider the problem of obtaining the best connected polygonal approximation. When does the original approach automatically yield a connected approximation?

5. What class of functions $h(r, s)$ possesses a representation of the form

$$h(r, s) = \min_{a,b} \int_r^s (u(t) - a - bt)^2 \, dt$$

for some function u?

6. Apply the dynamic programming approach to treat "nearest neighbor" variational problems of the form

$$\min_{0 \leqslant t_1 \leqslant t_2 \leqslant \cdots \leqslant t_N \leqslant T} [g(0, t_1) + g(t_1, t_2) + \cdots + g(t_N, T)].$$

7. Extend this method to consider the minimization of

$$g(0, t_1, t_2) + g(t_1, t_2, t_3) + \cdots + g(t_{N-1}, t_N, T).$$

8. Let $u(t)$ be a function defined in the interval $[0, T]$ with the following properties:

   a. $u(t)$ is a cubic polynomial in the intervals $[t_1, t_2]$, $[t_2, t_3]$,..., $[t_{N-1}, t_N]$ where $0 < t_1 < t_2 < \cdots < t_N < T$,
   b. $u(t)$ is linear in $[0, t_1]$ and $[t_N, T]$,
   c. $u(t_i) = b_i$, $i = 1, 2,..., N$,
   d. $u'(t)$ is continuous.

Then $u(t)$ exists and is unique and minimizes $J(u) = \int_0^T u''^2\, dt$ over functions satisfying conditions a, b, c, and d.

9. Consider a variant in which $u'(t_i)$ has a fixed value y and the aim is to minimize $\int_{t_i}^T u''^2\, dt$. Let $p(b_i, b_{i+1}, z, y)$ denote the cubic polynomial over $[t_i, t_{i+1}]$ determined by the conditions

$$p(t_i) = b_i, \quad p(t_{i+1}) = b_{i+1}, \qquad p'(t_i) = y, \quad p'(t_{i+1}) = z$$

and let $g(b_i, b_{i+1}, z, y) = \int_{t_i}^{t_{i+1}} p''^2\, dt$. Write

$$f_i(y) = \min J(y).$$

Then $f_i(y) = \min_z[g(b_i, b_{i+1}, z, y) + f_{i+1}(z)]$.

10. Show inductively that each $f_i(y)$ is quadratic in y, $f_i(y) = u_i + 2v_i y + w_i y^2$ and obtain recurrence relations for the u_i, v_i, and w_i, as functions of i.

11. Minimize $f_1(y)$ over y and thus solve the original problem. See

R. Bellman, B. Kashef, and R. Vasudevan, "Dynamic Programming and Splines," *J. Math. Anal. Appl.* (to appear).

12.29. Partial Differential Equations

Our application of dynamic programming to the study of the minimization of the functional

$$J(u) = \int_0^T g(u, u')\, dt \tag{12.29.1}$$

depended critically upon the simple decomposition

$$[0, T] = [0, S] + [S, T - S], \qquad 0 \leqslant S \leqslant T. \tag{12.29.2}$$

Similarly, we can use dynamic programming to study the minimization of functionals of the form

$$\int_R g(u, u_x, u_y)\, dA \tag{12.29.3}$$

using an analogous decomposition of the form

$$R = S + R - S, \qquad S \in R. \tag{12.29.4}$$

Thus, for example, we can take R to be a rectangle and S a subrectangle (Fig. 12.8) or R a triangle and S a trapezoid (Fig. 12.9).

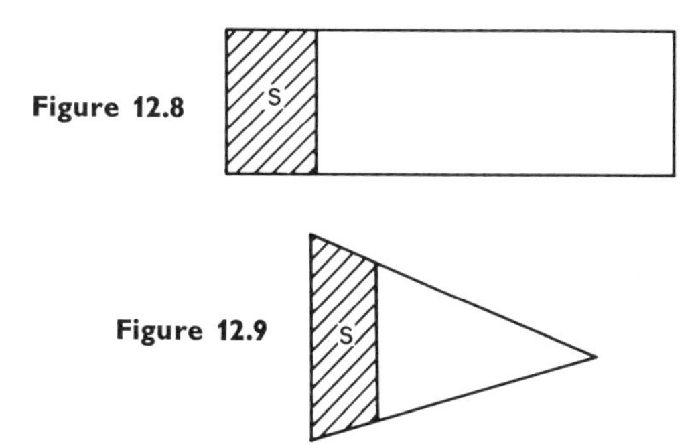

Figure 12.8

Figure 12.9

We can also employ more general figures (Fig. 12.10). By this means we can readily treat quadratic functionals and their associated linear partial differential equations, e.g., the Dirichlet functional

$$D(u) = \int_R (u_x^2 + u_y^2)\, dA \qquad (12.29.5)$$

and the potential equation

$$u_{xx} + u_{yy} = 0. \qquad (12.29.6)$$

A discussion of more general functionals would entail the study of partial differential functional equations, as indicated in the exercises.

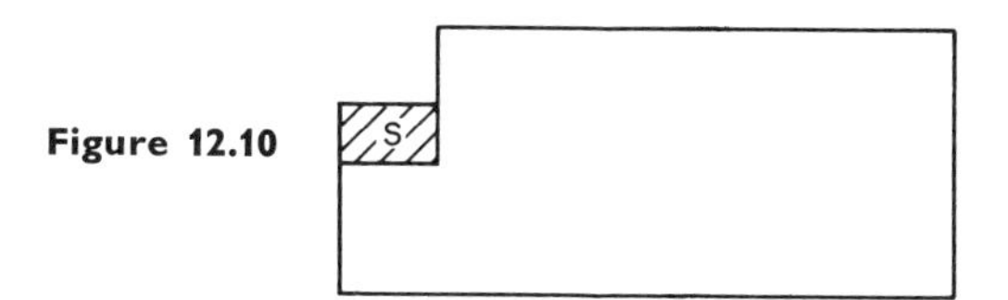

Figure 12.10

12.30. Successive Approximations to Reduce Dimensionality

The standard technique in analysis to treat an obdurate functional equation

$$T(u) = 0 \qquad (12.30.1)$$

is to imbed it within a family of equations

$$T_N(u) = 0 \qquad (12.30.2)$$

(where N may be a continuous parameter) with two vital stability properties

(a) $T_N(u) \to T(u)$ as $N \to \infty$,
(b) The solutions of $T_N(u) = 0$ approach the solutions of $T(u)$ as $N \to \infty$, (12.30.3)

and finally, a vital feasibility property:

The equation in (12.30.2) is amenable to existing analytic and computational algorithms. (12.30.4)

Up until recently this meant effectively that $T_N(u)$ was a linear transformation. The advent of the digital computer has changed this drastically. The fundamental condition now is not so much linearity as dimensionality. We possess effective techniques for finding the computational solution of nonlinear ordinary and partial differential equations of moderate dimension. Hence, we now have enormously more freedom in our choice of approximating equations.

In the references there will be found some illustrations of this.

Miscellaneous Exercises

1. Consider the functional equation

$$f(v) = \max_u (2uv - f(u)).$$

Show that if $f(u)$ is strictly convex in u, then $f'(u)/2$ is a solution of $\varphi(\varphi(u)) = u$.

2. Show that if φ is a solution of the foregoing functional equation, then so is $g^{-1}(\varphi(g))$ for any g.

3. How many entire solutions of $\varphi(\varphi(u)) = u$ are there, subject to $\varphi(0) = 0$? For discussions of functional equations of this nature, see

A. F. Timan, *Trans. Acad. Sci. USSR* (*Math. Ser. IAN*), Vol. 29, 1964, pp. 35–47.)

4. Let f be a given function and let $\{\varphi_i\}$, $i = 1, 2, \ldots$, be a sequence of given functions. Set

$$F_N(f) = \min_{a_k} \left\| f - \sum_{k=1}^{N} a_k \varphi_k \right\|.$$

Show that $F_N(f) = \min_{a_N} F_{N-1}(f - a_N \varphi_N)$.

Consider the case where $\| \cdots \| = \int_0^T | \cdots |^2 \, dt$ so that

$$F_N(f) = \int_0^T \int k_N(s, t) f(s) f(t) \, ds \, dt$$

and obtain a recurrence relation for the sequence $\{k_j\}$.

5. Consider

$$f(c, T, k) = \min_u \left[\int_0^T (u'^2 + u^2 + ku^4)\,dt\right],$$

$u(0) = c$. Show that

$$f(c, T, k) = c^2 f(1, T, kc^2).$$

Write $f(c, T, k) = f_0(c, T) + kf_1(c, T) + \cdots$ and determine f_0 and f_1. Does the series in k have a nonzero radius of convergence?

6. Consider the problem of minimizing

$$J(u) = \int_a^T (u'^2 + \varphi(t)\,u^2)\,dt$$

over the class of functions u such that u' is piecewise constant, $u'(t) = u_k$, $t_{k-1} \leqslant t \leqslant t_k$, $k = 1, 2, \ldots, N$, $t_0 = a$, $t_N = T$. Write

$$\min_{\{u_i\}} J(u) = f_N(c, a).$$

Show that

$$f_N(c, a) = \min_{u_1} \left[\int_a^{t_1} [u_1^2 + \varphi(t)(c + u_1 t)^2]\,dt + f_{N-1}(c + u_1 t_1)\right],$$

$N \geqslant 1$, with

$$f_1(c, a) = \min_{u_1} \left[\int_a^T [u_1^2 + \varphi(t)(c + u_1 t)^2]\,dt\right].$$

Show that $f_N(c, a) = r_N(a)c^2$ and thus obtain recurrence relations for the $r_N(a)$.

7. Let

$$f(c, T) = \min_u \int_0^T g(u, u')\,dt, \qquad u(0) = c,$$

$$F(c, T) = \min_u \int_0^T G(u, u')\,dt, \qquad u(0) = c.$$

Let $v(c, T)$ and $w(c, T)$ be the minimizing functions respectively in

$$f_T = \min_v [g(c, v) + vf_c], \qquad f(c, 0) = 0,$$

$$F_T = \min_w [G(c, w) + wF_c], \qquad F(c, 0) = 0.$$

Then

$$(f - F)_T \leqslant g(c, w) - G(c, w) + w(f - F)_c\,,$$
$$(F - f)_T \leqslant G(c, v) - g(c, v) + v(f - F)_c\,.$$

What conclusion can we draw about uniqueness of solution of the equation for f ?

8. Consider the problem of minimizing

$$J(x, y) = x_1(T)^2 + \int_0^T (y, y)\, dt,$$

with respect to y where $x' = Ax + y$, $x(0) = c$, and $x_1(T)$ denotes the first component of $x(T)$. Show that the minimization can be carried out in terms of functions of one variable. Hint: Solve for x and write

$$x = e^{At}c + \int_0^t e^{A(t-t_1)} y(t_1)\, dt_1\,.$$

Then

$$J(x, y) = \left(\sum_{k=1}^{N} a_k c_k + \int_0^T \left(\sum_{k=1}^{N} b_k(t_1)\, y_k(t_1)\right) dt_1\right)^2 + \int_0^T (y, y)\, dt,$$

where the a_k and $b_k(t)$ are known quantities. Start over with the problem of minimizing

$$J_a(y) = \left(b + \int_a^T (b(t), y)\, dt\right)^2 + \int_0^T (y, y)\, dt,$$

where $-\infty < b < \infty$, $0 \leqslant a \leqslant T$. Write

$$\varphi(b, a) = \min_y J_a(y),$$

show that $\varphi(b, a)$ is quadratic in b, and derive Riccati equations for the coefficients as functions of a. See

R. Bellman, *Introduction to the Mathematical Theory of Control Processes*, Academic Press, New York, 1967, p. 199.

9. Apply the same idea to the minimization of

$$J(u, v) = \sum_{k=1}^{R} a_k u(x_k\,, T)^2 + \int_0^T \int_0^1 v^2(x, t)\, dx\, dt,$$

where $u_t = u_{xx} + v$, $u(x, 0) = g(x)$, $u(0, t) = u(1, T) = 0$.

10. Apply the same idea to the minimization of

$$J(u, v) = u(T)^2 + \int_0^T v^2 \, dt,$$

where $u'(t) = a_1 u(t) + a_2 u(t-1) + v(t)$, $t \geqslant 1$, with $u(t) = g(t)$, $0 \leqslant t \leqslant 1$.

11. Discuss how the foregoing techniques may be used to provide a reduction of dimensionality for general terminal control processes of the form described by

$$J(x, y) = g(x_1(T), x_2(T), \ldots, x_R(T)) + \int_0^T h(y) \, dt,$$

where $x' = h(x, y)$, $x(0) = c$, using successive approximations.

12. Consider the minimization of the functional

$$J(u) = \int_S^T g(u, u', t) \, dt$$

subject to $u(S) = a$, $u(T) = b$. Assume that the problem is well-defined for $-\infty < a$, $b < \infty$, $0 < S < T < \infty$. Write $f(a, b; S, T) = \min_u J(u)$. Then $f(a, b, S, T) = \min_c [f(a, c; S, R) + f(c, b; R, T)]$, $S < R < T$.

13. Suppose that $g = u'^2 + k(t)u^2 + 2h(t)u$. Show that

$$f(0, 0, S, T) = \int_S^T \int_S^T k(t, t_1, T) \, h(t) \, h(t_1) \, dt \, dt_1$$

and use the foregoing result to obtain a functional equation for the Green's function.

14. If $h = 0$, then $f(a, b; S, T) = a^2 r_{11}(S, T) + 2ab r_{12}(S, T) + b^2 r_{22}(S, T)$. Use the foregoing result to obtain functional equations for the $r_{ij}(S, T)$.

15. Use the result in Exercise 12 to calculate $r_{ij}(0, 2^N)$ in terms of $r_{ij}(0, 2^{N-1})$. Discuss the computational efficacy of this as opposed to the use of the Riccati equation.

16. In the case where $g = u'^2 + \sum_{n=1}^{\infty} k_n(t) u^{2n}$ and

$$f(a, b; S, T) = \sum_{m+n \geqslant 2} a_m b^n r_{mn}(S, T),$$

obtain the corresponding functional equations for the $r_{mn}(S, T)$.

17. Obtain corresponding results for the multidimensional case. The foregoing is taken from

    R. Bellman, "Functional Equations in the Theory of Dynamic Programming—XVIII: Minimum Convolutions and Green's Functions," *J. Math. Anal. Appl.* (to appear).

    For a derivation of the Hadamard variational formula, see

    R. Bellman and H. Osborn, "Dynamic Programming and the Variation of Green's Functions," *J. Math. Mech.*, Vol. 7, 1958, pp. 81–86.

    There are some very interesting problems connected with the representation of the minimum of the Dirichlet functional.

18. Given that the function $v(c, t)$ is the optimal policy for the minimization of $J(u) = \int_0^T g(u, u')\, dt$, $u(0) = c$, determine the set of possible g's. This is a particular case of the "inverse problem in the calculus of variations." See

    O. Bolza, *Vorlesungen uber Variationsrechnung*, Koehler und Amelong, Leipzig, 1941.

    R. Bellman, "Dynamic Programming and Inverse Optimal Problems in Mathematical Economics," *J. Math. Anal. Appl.*, Vol. 29, 1970, pp. 424–428.

    E. Tonti, *Variational Principles*, Tamburini, Editori, Milano, 1968.

19. Starting with the equation $f_T = \min(g(c, v) + vf_c)$, show that we are led to $(gg - vg_v)_c = (-g_v)_T$, which, in principle, determines g.

20. Alternatively, show that f satisfies $(f_T)_T - v(f_T)_c = 0$. What are advantages and disadvantages of the two approaches?

21. Discuss the cases: a. $v = v_1(c)$, b. $v = v_2(T)$, c. (similarity solution), $v = v_3(k(c)h(T))$.

22. Consider the multidimensional case.

23. Let $A_N = (a_{n-m})$, $n, m = 1, 2, \ldots, N$ be a Toeplitz matrix and consider the minimization of $(x, A_N x) - 2(b, x)$ using dynamic programming. Does this procedure have any advantages over the usual procedures for solving $A_N x = b$?

24. Consider the function equation

$$f(x) + g(p) = \max_{0 \leqslant y \leqslant x} [pf(x + y) + (1 - p)f(x - y)],\ 0 < p \leqslant 1/2$$

    What are possible solutions? See

    R. Bellman, "Functional Equations in the Theory of Dynamic Programming, "*Mathematical Biosciences*, Vol. II, 1971, pp. 1–3.

25. Consider the functional equation $f_i = \min_{j \neq i}[t_{ij} + f_j]$, $i = 1, 2, \ldots, N$, $f_N = 0$ which occurs in routing problems, supposing thet $t_{ij} \geqslant 0$. Establish existence and uniqueness theorems. See

R. Bellman, K. L. Cooke and J. Lockett, *Algorithms, Graphs and Computers*, Academic Press, New York, 1970.

B. A. Carre, "An Algorithm for Network Routing Problems," *J. Inst. Math. Appl.*, 1971, pp. 273–294.

26. Show that $u_n \geqslant 0$, $u_{m+n} \leqslant u_m + u_n$ implies that u_N/N has a limit as $N \to \infty$. (Fekete–Polya–Szego).

27. Consider $f_n(p) = \max_q[b(p, q) + f_{n-1}(T(p, q))]$, $N \geqslant 1$, $f_0(p)$ given, $b(p, q) \geqslant 0$. Under what conditions can we assert that $f_N(p) \sim Na$, a independent of p? See

R. Bellman, "Functional Equations in the Theory of Dynamic Programming—XI: Limit Theorems," *Rend. Circ. Mat. Palermo*, Serie 11 – Tom VIII, 1959, pp. 1–3.

28. Consider the functional equation

$$f(p) = \inf_i \left\{ \sum_{j=1}^{N} p_j f(s_j) + b_i f(T_i p) + a_i \right\}$$

where $p = (p_1, p_2, \ldots, p_N)$. Under what conditions on the transformations T_i and the scalars a_i does it possess a unique bounded solution? See

R. Bellman and T. Brown, "Projective Metrics in Dynamic Programming," *Bull. Amer. Math. Soc.*, Vol. 71, 1965, pp. 773–775.

Further references to the powerful method of projective metrics developed by G. Birkhoff will be found there.

29. Consider the two-point boundary value problems,

$$(p(t)\, u')' + q(t)u = 0, \qquad u(a) = 1, \quad u(1) + bu'(1) = 0,$$

and the associated problem of maximizing the quadratic functional

$$J(u, v) = \int_a^1 (q(t)\, u^2 - p(t)\, u'^2 - 2uv(t))\, dt - \frac{p(t)}{b}\, u(1)^2.$$

Using a dynamic programming approach, obtain an equation for the variation of the associated Green's functions as a varies.

30. Replacing $q(t)$ by $\lambda + q(t)$ obtain thereby equations for the variation of the characteristic values and characteristic functions.

31. Obtain corresponding results for the vector–matrix case. See

R. Bellman and S. Lehman, "Functional Equations in the Theory of Dynamic Programming–X: Resolvents, Characteristic Functions and Values," *Duke Math. J.*, Vol. 27, 1960, pp. 55–70.

For another approach, see

K. S. Miller and M. M. Schiffer, "On the Green's Functions of Ordinary Differential Systems," *Proc. Amer. Math. Soc.*, Vol. 3, 1952, pp. 433–441.

Bibliography and Comments

§12.1. For more extensive discussion, application, and references, see the books

R. Bellman, *Dynamic Programming*, Princeton Univ. Press, Princeton, New Jersey, 1957.

R. Bellman and S. Dreyfus, *Applied Dynamic Programming*, Princeton Univ. Press, Princeton, New Jersey, 1962.

R. Bellman, *Adaptive Control Processes: A Guided Tour*, Princeton Univ. Press, Princeton, New Jersey, 1961.

§12.11. For further discussion, see

S. E. Dreyfus, *Dynamic Programming and the Calculus of Variations*, Academic Press, New York, 1965.

S. E. Dreyfus, "Dynamic Programming and the Hamilton-Jacobi Method of Classical Mechanics," *J. Opt. Theory Appl.*, Vol. 2, 1968, pp. 15–27.

A large number of references to interconnections between the classical calculus of variations, the maximum principle of Pontrjagin and dynamic programming will be found in

R. Bellman, *Introduction to Modern Control Theory*, Vol. 2, Nonlinear Processes, Academic Press, New York, 1970.

See also

E. B. Lee and L. Markus, *Foundations of Optimal Control Theory*, Wiley, New York, 1968.

V. G. Boltyanski, "Sufficient Conditions for Optimality and Justification of the Dynamic Programming Method," *SIAM J. Control*, Vol. 4, 1966, pp. 326–361.

§12.12. The nonlinear partial differential equation in (12.12.7) is variously known as the Hamilton–Jacobi–Bellman equation, or the Bellman equation. See

A. T. Fuller, "Optimization of some Nonlinear Control Systems by Means of Bellman's Equation and Dimensional Analysis," *Int. J. Control*, Vol. 3, 1966, pp. 359–394.

I. V. Girsanov, "Certain Relations Between the Bellman and Krotov Functions for Dynamic Programming Problems," *J. SIAM Control*, Vol. 7, 1969, pp. 64–67.

V. G. Pavlov and V. P. Cheprasov, "Constructing Certain Invariant Solutions of Bellman's Equation," *Automat. Remote Control*, January, 1968, pp. 31–36.

S. B. Gershwin, "On the Higher Derivatives of Bellman's Equation," *J. Math. Anal. Appl.*, Vol. 27, 1969.

§12.14. This solution of the nonlinear partial differential equation was first indicated in

R. Bellman, "On a Class of Variational Problems," *Quart. Appl. Math.*, Vol. 14, 1957, pp. 353–359.

This marks the first appearance of the Riccati equation in control theory and dynamic programming. Further developments were made by Adorno, Beckwith, Freimer, and Kalman. In particular, stochastic versions of processes of this nature lead to the Kalman–Bucy filter. See

H. Cox, "On the Estimation of State Variables and Parameters for Noisy Dynamic Systems," *IEEE Trans. Auto. Control*, Vol. AC–9, 1964, pp. 5–12.

One of the important questions of analysis is that of the variation of Green's functions as the region varies. This was first studied by Hadamard, the celebrated Hadamard variational formula. It can be obtained by dynamic programming techniques, cf.

R. Bellman and H. Osborn, "Dynamic Programming and the Variation of Green's Functions," *J. Math. Mech.*, Vol. 7, 1958, pp. 81–86.

and by imbedding, cf.

J. Devooght, "Variation of Green's Functions," *J. Math. Phys.*, Vol. 7, 1966, pp. 1764–1770.

§12.16. See

M. Aoki, "Note on Aggregation and Bounds for the Solution of the Matrix Riccati Equation," *Math. Anal. Appl.*, Vol. 21, 1968, pp. 377–383.

§12.19. There are a number of interesting analytic questions connected with functional equations of this type; see

R. Bellman and K. Cooke, "Existence and Uniqueness Theorems in Invariant Imbedding—II: Convergence of a New Difference Algorithm," *J. Math. Anal. Appl.*, Vol. 12, 1965, pp. 246–253.

See the discussion of "soft solutions" in

W. F. Noh and M. H. Protter, "Difference Methods and the Equations of Hydrodynamics," *J. Math. Mech.*, Vol. 12, 1963, pp. 149–192.

12.21. See also

N. Froman and P. O. Froman, *JWKB Approximation, Contributions to the Theory*, North–Holland, Publ., Amsterdam, 1965.

§12.23. See

R. T. Rockafellar, "Generalized Hamiltonian Equations for Convex Problems of Lagrange," *Pacific J. Math.*, Vol. 33, 1970, pp. 411–427.

§12.24. See

R. Bellman, "Functional Equations in the Theory of Dynamic Programming—XV: Layered Functionals and Partial Differential Equations," *J. Math. Anal. Appl.*, Vol. 28, 1969, pp. 1–3.

For an extensive application of this approach, see

E. Hopf, "Generalized Solutions of Nonlinear Equations of the First Order," *J. Math. Mech.*, Vol. 14, 1965, pp. 951–972.

The method was first used by Bellman and Lax to handle the equation $u_t = g(u_x)$; see the discussion in

R. Kalaba, "On Nonlinear Differential Equations, the Maximum Operation, and Monotone Convergence," *J. Math. Mech.*, Vol. 8, 1959, pp. 519–574; especially p. 567.

§12.25. For a discussion of stochastic control processes and the ways in which nonlinear partial differential equations enter, see

R. L. Stratonovich, *Conditional Markov Processes and Their Application to the Theory of Optimal Control*, American Elsevier, New York, 1968.
A. A. Fel'dbaum, *Optimal Control Systems*, Academic Press, New York, 1965.
H. J. Kushner, *Stochastic Stability and Control*, Academic Press, New York, 1967.
W. Fleming, "Duality and a priori Estimates in Markovian Optimization Problems," *J. Math. Anal. Appl.*, Vol. 16, 1966, pp. 254–279.
W. Fleming, "Some Markovian Optimization Problems," *J. Math. Mech.*, Vol. 12, 1963, pp. 131–140.
H. J. Kushner, *Methods for the Numerical Solution of Degenerate Linear and Nonlinear Elliptic Boundary Value Problems*, Brown Univ., Providence, Rhode Island, TR 67–7, 1967.

§12.26. For classical semigroup theory, see

E. Hille and R. Philips, *Functional Analysis and Semi-Groups*, *Collog. Publ. Amer. Math. Soc.*, 1948; 1957.

See also

F. E. Browder, "On the Unification of the Calculus of Variations and the Theory of Monotone Nonlinear Operators in Banach Spaces," *Proc. Nat. Acad. Sci. USA*, Vol. 56, 1966, pp. 419–425,

and a number of subsequent papers by the same author; cf. also

G. J. Minty, *Proc. Nat. Acad. Sci. USA*, Vol. 50, 1963, pp. 1038–1041

for the theory of monotone operator equations.

§12.27. See

R. Bellman, R. Kalaba, and J. Lockett, *Numerical Inversion of the Laplace Transform*, American Elsevier, New York, 1966.

§12.28. See

R. Bellman, "On the Approximation of Curves by Line Segments using Dynamic Programming," *Comm. ACM*, Vol. 4, 1961, p. 284.
R. Bellman, "Curve Fitting By Segmented Straight Lines," *J. Amer. Stat. Assoc.*, Vol. 64, 1969, pp. 1079–1084.
M. Roth and R. Roth, "Segmental Differential Approximations and Biological Systems: An Analysis of a Metabolic Process," *J. Theoret. Biol.*, Vol. 11, 1966, pp. 168–176.
R. Bellman and R. Roth, "A Technique for the Analysis of a Broad Class of Biological Systems," *Bionics Symp.*, 1966.
R. Bellman, B. Gluss, and R. Roth, "Segmental Differential Approximation and the 'Black Box' Problem," *J. Math. Anal. Appl.*, Vol. 12, No. 1, 1965, pp. 91–104.
R. Bellman, B. Gluss, and R. Roth, "On the Identification of Systems and the Unscrambling of Data: Some Problems Suggested by Neurophysiology," *Proc. Nat. Acad. Sci.*, Vol. 52, 1964, pp. 1239–1249.

§12.29. See

E. Angel and R. Bellman, *Partial Differential Equations and Dynamic Programming*, Academic Press, New York, 1972.

§12.30. The problem of dimensionality is a major one in dynamic programming to which a great deal of effort has been directed. See

A. O. Esogbue and A. J. Singh, *Reduction of Dimensionality in Dynamic Programming of Higher Dimensions: A Comparative Study and Analysis of Computational Aspects*, Tech. Memo. 237, Operations Res. Dept., Case Western Reserve Univ., July 1971.

A. Lew, "Reduction of Dimensionality by Approximation Techniques: Diffusion Processes," *J. Math. Anal. Appl.*, Vol. 37, 1972.

R. E. Bellman and S. E. Dreyfus, *Applied Dynamic Programming*, Princeton Univ. Press, Princeton, New Jersey, 1961.

D. C. Collins, *Reduction of Dimensionality in Dynamic Programming via the Method of Diagonal Decomposition*," University of Southern California, Los Angeles, August 1969.

D. C. Collins, *Terminal State Dynamic Programming*, Tech. Rep. USCEE, Univ. of Southern California, 1969.

A. E. Durling, *Computational Aspect of Dynamic Programming in Higher Dimensions*, Tech. Rep. TR-64-3, Electrical Eng. Depart., Syracuse Univ., Syracuse, New York, May 1964.

R. E. Larson, *State Increment Dynamic Programming*, American Elsevier, New York, 1968.

P. J. Wong and D. G. Luenberger, "Reducing the Memory Requirement of Dynamic Programming," *Operations Res.*, Vol. 16, 1968, pp. 1115–1125.

P. J. Wong, "A New Decomposition Procedure for Dynamic Programming," *Operations Res.*, Vol. 18, 1970, pp. 119–131.

Chapter 13 INVARIANT IMBEDDING

13.1. Introduction

In this chapter we wish to present some of the basic ideas of a new approach to many areas of analysis and mathematical physics, the theory of invariant imbedding. In addition, we will give some applications, beginning with general two-point boundary value problems not necessarily of variational origin. As we shall see, linear equations give rise to Riccati equations, while nonlinear equations generate nonlinear partial differential equations, subject, however, to initial conditions, a fact which renders them of immediate computational value.

Then we continue the discussion of the asymptotic behavior of the solutions of nonlinear systems of the form

$$\frac{dx}{dt} = Ax + g(x), \qquad x(0) = c, \tag{13.1.1}$$

begun in Chapter 3, Volume I, and complete it in certain details. Further results, however, will be given in Chapter 14. We then turn to linear equations of the form

$$u'' + (1 + f(t))u = 0, \tag{13.1.2}$$

using similar ideas. Finally we shed further light on the significance of the WKB approximation for the solution of

$$u'' + a(t)u = 0, \tag{13.1.3}$$

by considerations of some simple aspects of wave propagation.

Extensive references will be given to further analytic and computational results, and to applications of the basic ideas of invariant imbedding to other physical processes.

Throughout we are constantly emphasizing the idea that a great deal of mathematical insight can be gained by associating an equation with a physical process and then using obvious properties of the process.

13.2. Maximum Altitude

In order to present the methodology of a different approach to the familiar processes of mathematical physics in a very simple setting, consider the problem of determining the maximum altitude achieved by a particle traveling straight upward in a vacuum with initial velocity v. Analytically, making the usual preliminary assumptions, we can obtain the solution by means of the differential equation

$$u'' = -g, \qquad u(0) = 0, \quad u'(0) = v. \tag{13.2.1}$$

The explicit solution is readily seen to be

$$u = vt - \frac{gt^2}{2}. \tag{13.2.2}$$

We see then that the maximum altitude is achieved at the time

$$t_{\max} = \frac{v}{g}, \tag{13.2.3}$$

while the maximum altitude is given by

$$u_{\max} = \frac{v^2}{2g}. \tag{13.2.4}$$

Observe that in order to obtain the desired information, namely the maximum altitude as a function of v, it was apparently necessary first to derive u as a function of t for $t \geqslant 0$. We take this conventional route as a matter of course, with the roundaboutness of the path forgiven because of the simple explicit answer we ultimately derive. Can we, however, answer the original question without this circumnavigation? In so doing, we may develop a method which will be far superior in other situations where the usual technique encounters serious obstacles.

13.3. Functional Equation Approach

To avoid the intermediary step of using the explicit structure of the solution as a function of time, we proceed in the following fashion. Let us introduce the function

$$f(v) = \text{the maximum altitude attained starting with initial velocity } v. \tag{13.3.1}$$

Here $v \geqslant 0$.

With reference to Fig. 13.1, we can assert that the maximum altitude, OR, attained starting with initial velocity v_0 at 0 is equal to OP, plus the maximum additional altitude gained starting at P with the initial velocity v_p, where v_p is the new velocity at p. Here p is any intermediate point.

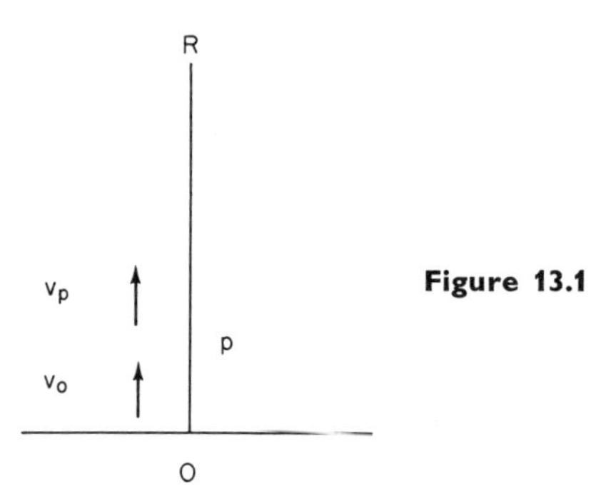

Figure 13.1

To obtain a simple equation for $f(v)$ using this property, we take OP to be the distance traveled in an infinitesimal time Δ. Then, to terms which are $0(\Delta^2)$, we have

$$OP = v\Delta, \qquad v_p = v - g\Delta. \tag{13.3.2}$$

Hence, we have the functional equation

$$f(v) = v\Delta + f(v - g\Delta) + O(\Delta^2). \tag{13.3.3}$$

Writing

$$f(v - g\Delta) = f(v) - g\Delta f'(v) + O(\Delta^2), \tag{13.3.4}$$

and letting $\Delta \to 0$, we obtain the differential equation

$$f'(v) = \frac{v}{g}, \tag{13.3.5}$$

with the initial condition $f(0) = 0$. Hence,

$$f(v) = \frac{v^2}{2g}, \tag{13.3.6}$$

agreeing with the result of (13.2.4).

Exercises

1. Derive the result (13.2.3) using the same reasoning.
2. What is required to make the foregoing derivation completely rigorous? Can it be supplied?

13.4. Invariant Imbedding

The classical approach to the problem discussed above is one of imbedding. The particular question of determining the maximum altitude is imbedded within the family of questions concerning the position of the particle at an arbitrary time t. The approach of the previous section is also one of imbedding, but one where the family of associated problems is quite different from that involving time. To indicate both of these aspects, the similarity and the difference, we coin the name "invariant imbedding."

There is a strong resemblance between the fundamental ideas of invariant imbedding and dynamic programming. Both emphasize the multistage aspect of processes and both concentrate upon observables. It was the dynamic programming methodology which inspired invariant imbedding, which is a simpler theory in many ways since the optimization aspects are missing.

13.5. Inhomogeneous Medium

If we formulate a corresponding question for the case where the atmosphere is considered to be an inhomogeneous medium, we face the problem of determining $u_{\max}$ for an equation of the form

$$u'' = g(u, u'), \qquad u(0) = h, \quad u'(0) = v. \tag{13.5.1}$$

An initial attempt to use the method of invariant imbedding will show us that we cannot manage with a function solely of v. We need, as is physically obvious, the function

$$f(h, v) = \text{the maximum additional altitude attained starting at altitude } h \text{ with upward velocity } v. \tag{13.5.2}$$

The same reasoning as before yields the functional equation

$$f(h, v) = \Delta v + f(h + v\Delta, v + \Delta g(h, v)) + O(\Delta^2), \tag{13.5.3}$$

and thus, in the limit, the partial differential equation

$$0 = v + v f_h + g(h, v) f_v , \tag{13.5.4}$$

with the initial condition

$$f(h, 0) = 0. \tag{13.5.5}$$

We suppose that $h, v \geqslant 0$. We will discuss some of the computational aspects below.

Exercises

1. Consider the case where $g(u, u') = -g - \epsilon u^2$. Obtain a perturbation expansion for $f(h, v)$.
2. Consider the case where $g(u, u') = -g - \epsilon(u')^k$. Obtain a perturbation expansion for $f(h, v)$. Do we need the function of two variables $f(h, v)$ in this case?
3. Determine $f(h, v)$ explicitly for $g(u, u') = -(u + u')$.
4. If f is to depend solely on v, what form must g have?

13.6. Computational Aspects

In order to solve (13.5.4) subject to (13.5.5) for a general function $g(u, u')$, it will be necessary to employ some type of numerical procedure. We can either use the theory of characteristics, use a conventional difference algorithm based on (13.5.4), or use an unconventional difference algorithm of the type suggested by (13.5.3). We have already covered a good deal of this ground in the preceding chapter. We will return to this topic below in Sec. 13.15.

The reader may wonder why we do not simply use a straightforward computational approach based upon (13.5.1), i.e., the ability to employ a digital computer to trace out the numerical solution of an initial-value problem with great ease and rapidity, rather than go to all of this effort merely to end up using a computer once again on a much more complex equation.

The answer resides in efficiency and time. Observe that the direct approach means that for each set of values of h and v, we have to tabulate the values of $u(t) = u(t, h, v)$ for $0 \leqslant t \leqslant T$ in order to determine $u_{\max}$ and $t_{\max}$; see Fig. 13.2. If the function $f(h, v)$ is desired for a wide

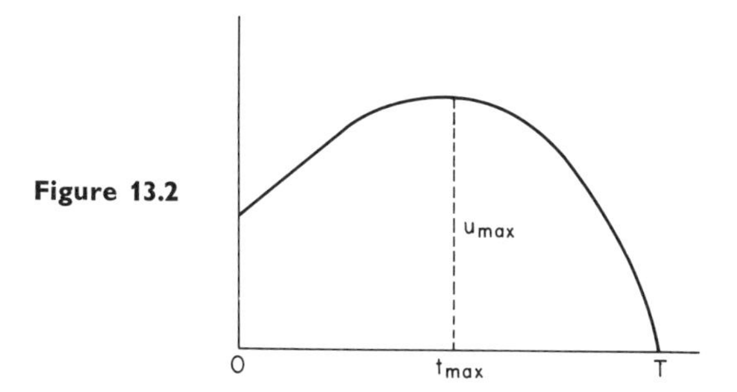

Figure 13.2

range of values of h and v, this can be a time-consuming endeavor. In any case, it is esthetically displeasing.

On the other hand, using (13.5.3), each value of $f(h, v)$ that is calculated is the answer to a particular question.

13.7. Two-point Boundary Value Problems

Consider next the problem of obtaining the solution of

$$u'' = g(u, u'), \tag{13.7.1}$$

subject to a two-point condition, say

$$u(0) = c_1, \qquad u(T) = c_2. \tag{13.7.2}$$

In the previous pages we approached this problem by means of two distinct methods, quasilinearization in Chapter 11 and dynamic programming in Chapter 12, albeit there only for the case where (13.7.1) was equivalent to an Euler equation. Let us now indicate how to employ invariant imbedding to treat (13.7.1) subject to (13.7.2). To this end, let us consider the more general equation

$$\begin{aligned} u' &= g(u, v), \qquad u(0) = c_1, \\ v' &= h(u, v), \qquad v(T) = c_2. \end{aligned} \tag{13.7.3}$$

We fix these particular conditions to fit the physical model we will use for intuitive purposes below. As a matter of fact, this process, a transport process, was the focus of much of the original work in invariant imbedding.

Examining (13.7.3), we see that a numerical solution can be readily effected if either $v(0)$ or $u(T)$ is obtained, since the system would now be subject to initial conditions. These missing values, $v(0)$ and $u(T)$, clearly depend on c_1, c_2, and T. Guided by the foregoing, let us note this explicitly,

$$v(0) = r(c_1, c_2, T), \qquad u(T) = s(c_1, c_2, T), \tag{13.7.4}$$

for $T \geqslant 0$, $-\infty < c_1, c_2 < \infty$. Can we obtain equations for these functions without the necessity for determining u and v as functions of t? For the moment we shall proceed formally, assuming that these functions are well defined. The general question of the domain of definition of these functions is one of great difficulty which we return to below.

Observe the difference in viewpoint. In place of considering c_1, c_2, and T as fixed parameters with the emphasis upon $u(t)$, $v(t)$, for $0 \leqslant t \leqslant T$, we regard c_1, c_2, and T as the fundamental parameters, with the attention upon the functions $r(c_1, c_2, T)$ and $s(c_1, c_2, T)$. As we shall see, they possess simple physical interpretations.

13.8. A Simple Transport Process

To provide a physical motivation for the mathematical analysis that follows, let us consider the following simple version of a transport process. Subsequently, we shall consider a multidimensional version. A stream of particles travels in both directions along a rod of finite length, as indicated in Fig. 13.3, subject to interactions we will describe in a moment.

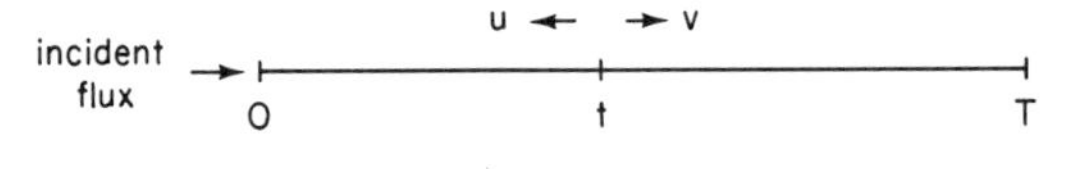

Figure 13.3

Let $u(t)$ denote the flux to the left at an internal point t, and $v(t)$ the flux to the right. Let us suppose that there is no interaction with the medium, but that interaction does occur between the opposing streams, which is to say, we have "collisions."

Let

$$\begin{aligned} \varphi(u, v)\Delta &= \text{rate of decrease of right-hand flux due to collision of opposing streams of intensities } u \text{ and } v \text{ respectively, in the interval } [t, t + \Delta], \qquad (13.8.1) \\ \psi(u, v)\Delta &= \text{rate of decrease of left-hand flux due to collision of opposing streams of intensities } u \text{ and } v \text{ respectively, in } [t - \Delta, t]. \end{aligned}$$

Then, to terms in Δ^2,

$$v(t+\Delta) = v(t) - \varphi(u, v)\Delta, \qquad u(t-\Delta) = u(t) - \psi(u, v)\Delta. \tag{13.8.2}$$

Hence, we obtain the system of differential equations

$$v'(t) = -\varphi(u, v), \qquad u'(t) = \psi(u, v). \tag{13.8.3}$$

If we assume an incident flux at 0, the left boundary, and none at T, the right boundary, as in Fig. 13.3, the boundary conditions are

$$v(0) = c, \qquad u(T) = 0. \tag{13.8.4}$$

We see then one physical interpretation of a system of differential equations of the type appearing in (13.7.3). What is interesting to note is that the physical process automatically introduces a two-point boundary condition.

13.9. Invariant Imbedding: Particle Counting

Let us now see how we can apply invariant imbedding to determine the missing initial conditions, $u(0)$ and $v(T)$. Let us reverse Fig. 13.3 to be consistent with notation used in the earlier work cited in the references.

The method we present first is called "particle counting" because it is based upon examination of the underlying physical process. Although it requires some auxiliary effort to be made rigorous, it has some important advantages. In particular, it allows us to make simplifying approximations readily based upon intuitive physical considerations. Let us replace T by a (Fig. 13.4). A rigorous derivation along conventional lines is given below in Sec. 13.10.

Introduce the function

$$r(c, a) = \text{the intensity of reflected flux from a rod of length } a \text{ due to incident flux of intensity } c. \tag{13.9.1}$$

This function is defined for $a \geq 0$, $c \geq 0$.

Figure 13.4

The incident flux of intensity c is diminished by interaction with the left-hand flux at $a - \Delta$. This diminished flux then represents an incident flux on a rod of length $(a - \Delta)$. Hence, the left-hand flux at $a - \Delta$ can be expressed in terms of the function defined in (13.9.1) with a replaced by $a - \Delta$. This left-hand flux is again diminished by interaction in $[a, a - \Delta]$ on the way out. Let us express all of this analytically.

We have

$$\begin{aligned} u(a - \Delta) &= r(c - \varphi(u(a - \Delta), c)\Delta, a - \Delta) \\ &= r(c - \varphi(u(a), c)\Delta, a - \Delta) + O(\Delta^2) \\ &= r(c - \varphi(r(c, a), c)\Delta, a - \Delta) + O(\Delta^2), \end{aligned} \tag{13.9.2}$$

and

$$r(c, a) = u(a) = u(a - \Delta) - \Psi(r(a, c), c)\Delta + O(\Delta^2). \tag{13.9.3}$$

Combining these last two equations, we have

$$r(c, a) = r(c - \varphi(r(c, a), c)\Delta, a - \Delta) - \Psi(r(a, c), c)\Delta + O(\Delta^2). \tag{13.9.4}$$

Passing to the limit as $\Delta \to 0$, we obtain the nonlinear partial differential equation

$$\frac{\partial r}{\partial a} = -\varphi(r, c)\frac{\partial r}{\partial c} - \Psi(r, c), \tag{13.9.5}$$

with the obvious initial condition

$$r(c, 0) = 0, \qquad c \geqslant 0. \tag{13.9.6}$$

Exercises

1. Determine the corresponding equation for $t(c, a)$, the transmitted flux.

13.10. Perturbation Analysis

In order to obtain a rigorous proof of the foregoing, it is necessary to impose conditions on the functions $g(u, v)$ and $h(u, v)$ which will ensure that the solution of (13.7.3) exists, is unique, and is a function of the parameters c_1, c_2, and a, possessing partial derivatives for c_1, c_2 and a in a specified region. It is not difficult to obtain results of this nature for small a under the usual assumptions that g and h satisfy Lipschitz conditions. The technique of quasilinearization of Chapter 11 can readily be used for this purpose. The general case where a is large is considerably more difficult and not much has been done in this direction.

Once we have established these results, a fairly straightforward perturbation analysis yields the result of (13.9.5) and corresponding results for more general boundary conditions.

To illustrate this approach, consider the equations

$$\begin{aligned} u'(t) &= g(u, v), & u(0) &= 0, \\ v'(t) &= h(u, v), & v(a) &= c, \quad 0 < t < a, \end{aligned} \tag{13.10.1}$$

and let U, V denote the solution of the corresponding equation for the interval $[0, a + \Delta]$,

$$\begin{aligned} U'(t) &= g(U, V), & U(0) &= 0, \\ V'(t) &= h(U, V\ , & V(a + \Delta) &= c, \quad 0 < t < a + \Delta. \end{aligned} \tag{13.10.2}$$

The equations in (13.10.2) can be converted into equations with new boundary conditions over $[0, a]$. We have

$$\begin{aligned} V(a) = V(a + \Delta) - \Delta V'(a) &= c - h(U(a), V(a))\Delta \\ &= c - h(u(a), v(a))\Delta, \end{aligned} \tag{13.10.3}$$

to terms in $0(\Delta^2)$. Again to this order, we introduce functions w and q by means of the relations

$$U = u + w\Delta, \qquad V = v + q\Delta, \tag{13.10.4}$$

for $0 \leqslant t \leqslant a$. Substituting in (13.10.2), we see that w and q satisfy the linear perturbation equations

$$\begin{aligned} w' &= wg_u + qg_v\,, & w(0) &= 0, \\ q' &= wh_u + qh_v\,, & q(a) &= -h(u(a), v(a)). \end{aligned} \tag{13.10.5}$$

Let us next turn to the linear perturbation equations for u_t and v_c. Here u_c, v_c denote the partial derivatives of the solution of (13.10.1) with respect to the parameter c. We have

$$\begin{aligned} u_c' &= u_c g_u + v_c g_v\,, & u_c(0) &= 0, \\ v_c' &= u_c h_u + v_c h_v\,, & v_c(a) &= 1. \end{aligned} \tag{13.10.6}$$

For small T, it is easy to demonstrate that this linear system possesses a unique solution. In any case, we assume that this has been demonstrated. By virtue of this uniqueness of solutions, (13.10.5) and (13.10.6) yield

$$w = -h(u(a), v(a))u_c\,, \qquad q = -h(u(a), v(a))v_c\,. \tag{13.10.7}$$

Finally, to $O(\Delta^2)$,

$$U(a + \Delta) = U(a) + \Delta U'(a) = U(a) + \Delta g(u, v)$$
$$= u(a) + w\Delta + \Delta g(u, v). \tag{13.10.8}$$

Setting

$$r(a, c) = u(a), \tag{13.10.9}$$

we see that (13.10.8) leads to

$$r(a + \Delta, c) = r(a, c) - \Delta h(r, c)u_c(a) + \Delta g(r, c), \tag{13.10.10}$$

upon using (13.10.7) and the condition $v(a) = c$. Hence, we have, in the limit as $\Delta \to 0$, the desired partial differential equation

$$\frac{\partial r}{\partial a} = -h(r, c)\frac{\partial r}{\partial c} + g(r, c). \tag{13.10.11}$$

Exercises

1. Set $s(a, c) = v(0)$. Show in the same fashion that

$$\frac{\partial s}{\partial a} = -h\,(r, t)\frac{\partial s}{\partial c}, \qquad s(0, c) = c.$$

2. Consider the case where $g(u, v) = a_{11}u + a_{12}v$, $h(u, v) = a_{21}u + a_{22}v$. Show that $r(a, c) = r(a)c$, $s(a, c) = s(a)c$, and obtain ordinary differential equations for $r(a)$ and $t(a)$ from (13.10.10) and Exercise 1.
3. Obtain the corresponding multidimensional results where

$$x' = g(x, y), \quad x(0) = 0, \qquad y' = h(x, y), \quad y(a) = c.$$

4. Obtain the corresponding results and initial conditions in a when the boundary conditions are $x(0) = c_1$, $y(a) = c_2$.
5. What results hold when the boundary conditions are $x(0) = c$, $y(a) = d$.
6. What results hold if $u' = g(u, v, t)$, $v' = h(u, v, t)$?
7. Solve for $u(t)$ and $v(t)$ given the functions $r(a)$ and $t(a)$.
8. Let the solution at an internal point t_1 be regarded as a function of c and a, say $z(c, a)$. Obtain an analogous equation for $z(c, a)$.

13.11. Linear Systems

Let us now turn to the important n-dimensional linear system

$$\begin{aligned} -x' &= Ax + Dy, \qquad x(a) = c, \\ y' &= Bx + Cy, \qquad y(0) = d \end{aligned} \tag{13.11.1}$$

and establish some results which are important for their own sake and which can be used to initate a study of the nonlinear system.

Take (X_1, Y_1) and (X_2, Y_2) to be the principal solutions of the matrix version of (13.11.1)

$$\begin{aligned} -X' &= AX + DY, \\ Y' &= BX + CY, \end{aligned} \tag{13.11.2}$$

i.e.

$$\begin{aligned} X_1(0) &= 1, \qquad Y_1(0) = 0, \\ X_2(0) &= 0, \qquad Y_2(0) = 1. \end{aligned} \tag{13.11.3}$$

We then find the solution of (13.11.1) in a familiar fashion, writing

$$x = X_1 b + X_2 d, \qquad y = Y_1 b + Y_2 d, \tag{13.11.4}$$

where the vector b is determined by the relation

$$c = X_1(a)b + X_2(a)d. \tag{13.11.5}$$

If $X_1(a)$ is nonsingular, b is uniquely determined. Since $X_1(0) = I$, it is clear that $X_1(a)^{-1}$ exists for small a. We wish to determine conditions under which it exists for all $a \geqslant 0$.

One way of doing this is to assume that the equations in (13.11.1) are the Euler equations of an associated variational problem. We discussed this case in Chapter 7, Volume I. Here we wish to accomplish this aim by identifying (13.11.1) with a linear transport process.

13.12. Linear Transport Process

Our basic physical process is an idealized transport process which may be described in the following terms. N different types of particles move in either direction along a line of finite length a. Although they do not interact with each other, they do interact with the medium. The effect of this interaction is to change a particle from one type to another, traveling in the same or the opposite direction.

Let us consider the passage of a particle of type i through the interval $[t, t + \Delta]$ where Δ is an infinitesimal, and i may assume any of the values 1, 2, ..., N. Let

$$
\begin{aligned}
a_{ij}(t)\Delta &= \text{the probability that a particle in state } j \text{ will be transformed into a particle in state } i \text{ traveling in the same direction, } j \neq i, \text{ upon traversing the interval } [t + \Delta, t] \text{ going to the right,} \\
1 - a_{ii}(t)\Delta &= \text{the probability that a particle in state } i \text{ will remain in state } i \text{ in the same direction while traversing the interval } [t + \Delta, t] \text{ going to the right,} \\
b_{ij}(t)\Delta &= \text{the probability that a particle in state } j \text{ will be transformed into a particle in state } i \text{ traveling in the opposite direction upon traversing the interval } [t + \Delta, t] \text{ going to the right.}
\end{aligned}
\tag{13.12.1}
$$

Similarly, we introduce the functions $c_{ij}(t)$ and $d_{ij}(t)$, associated with forward scattering and backward scattering for a particle going to the left through $(t, t + \Delta]$. We suppose that all of these functions are nonnegative for $t \geqslant 0$. The most important case is that where they are piecewise continuous. Suppose that there are fluxes incident at both ends of the line, as indicated in Fig. 13.1, of constant intensities per unit time. Let us now introduce the steady-state functions for $i = 1, 2, \ldots, N$ (Fig. 13.5),

$$
\begin{aligned}
x_i(t) &= \text{the expected flux to the right at the point } t, \\
y_i(t) &= \text{the expected flux to the left at the point } t,
\end{aligned}
\tag{13.12.2}
$$

where $0 \leqslant t \leqslant a$. We shall omit the term "expected" in what follows and proceed formally to obtain some equations for x_i and y_i. There is no need for rigorous details here, since we are using the stochastic process solely as a guide to our intuition in ascertaining some of the properties of the solution of (13.1.1).

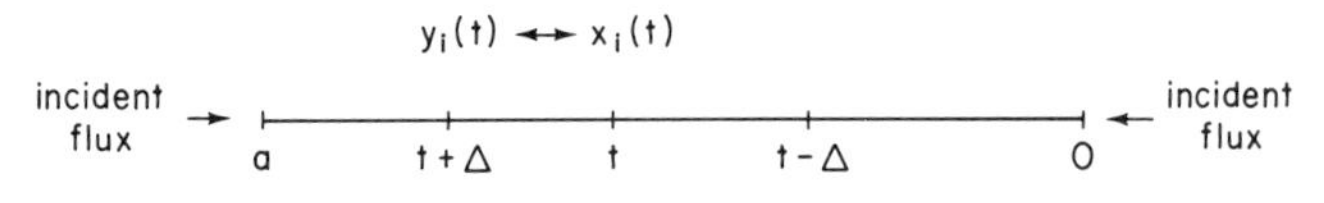

Figure 13.5

An input–output analysis (which is to say, a local application of conservation relations) of the intervals $[t - \Delta, t]$ and $[t, t + \Delta]$ yields the following relations, valid to $o(\Delta)$:

$$
\begin{aligned}
x_i(t) &= x_i(t + \Delta)(1 - a_{ii}\Delta) + \sum_{j \neq i} a_{ij}\, \Delta x_j(t + \Delta) + \sum_{j=1}^{N} d_{ij}\, \Delta y_j(t), \\
y_i(t) &= y_i(t - \Delta)(1 - c_{ii}\Delta) + \sum_{j \neq i} c_{ij}\, \Delta y_i(t - \Delta) + \sum_{j=1}^{N} b_{ij}\, \Delta x_j(t).
\end{aligned} \tag{13.12.3}
$$

Passing to the limit as $\Delta \to 0$, we obtain the differential equations

$$
\begin{aligned}
-x_i'(t) &= -a_{ii}x_i(t) + \sum_{j \neq i} a_{ij}x_j(t) + \sum_{j=1}^{N} d_{ij}y_j(t), \\
y_i'(t) &= -c_{ii}y_i(t) + \sum_{j \neq i} c_{ij}y_j(t) + \sum_{j=1}^{N} b_{ij}x_j(t),
\end{aligned} \tag{13.12.4}
$$

$i = 1, 2, \ldots, N$. Referring to Fig. 13.1, we see that the boundary conditions are

$$
x_i(a) = c_i, \qquad y_i(0) = d_i, \qquad i = 1, 2, \ldots, N. \tag{13.12.5}
$$

We introduce the matrices

$$
B = (b_{ij}), \qquad D = (d_{ij}),
$$

$$
A = \begin{pmatrix} -a_{11} & a_{12} & \cdots & a_{1N} \\ a_{21} & -a_{22} & \cdots & a_{2N} \\ \vdots & & & \\ a_{N1} & a_{N2} & \cdots & -a_{NN} \end{pmatrix}, \qquad C = \begin{pmatrix} -c_{11} & c_{12} & \cdots & c_{1N} \\ c_{21} & -c_{22} & \cdots & c_{2N} \\ \vdots & & & \\ c_{N1} & c_{N2} & \cdots & -c_{NN} \end{pmatrix}. \tag{13.12.6}
$$

B and D are nonnegative matrices, while A and C possess quite analogous properties despite the negative main diagonals. The conditions of pure scattering imposes conditions on the column sums of $A + B$ and $C + D$ which we will use below.

We observe then the special structure of the matrices A, B, C, D required for (13.1.1) to be associated with a linear transport process. Introducing the vectors x and y, with components x_i and y_i respectively, we obtain a linear vector-matrix system subject to a two-point boundary condition.

If no particles are lost in the process, we say that it is pure scattering; if particles disappear, we say absorption occurs. We will consider the pure scattering case first. We will not discuss the fission case where a critical length exists.

13.13. Reflection and Transmission Matrices

Returning to Sec. 13.11 and assuming that $d = 0$, we see that

$$x(t) = X_1(t)X_1(a)^{-1}c, \qquad y(t) = Y_1(t)X_1(a)^{-1}c. \tag{13.13.1}$$

The reflected and transmitted fluxes are given respectively by

$$y(a) = Y_1(a)X_1(a)^{-1}c, \qquad x(0) = X_1(a)^{-1}c. \tag{13.13.2}$$

These relations define reflection and transmission matrices

$$R(a) = Y_1(a)X_1(a)^{-1}, \qquad T(a) = X_1(a)^{-1} \tag{13.13.3}$$

which play a central role in what follows.

13.14. Invariant Imbedding

Let us now present a different analytic formulation of the scattering process which will permit us to demonstrate the existence of the reflection and transmission matrices for all $a \geqslant 0$. To begin with, we introduce the function

$$r_{ij}(a) = \text{the reflected flux in state } i \text{ from rod of length } a \text{ due to an incident flux of unit intensity in state } j. \tag{13.14.1}$$

This is defined for $a \geqslant 0$, $i, j = 1, 2, \ldots, N$. See Fig. 13.6.

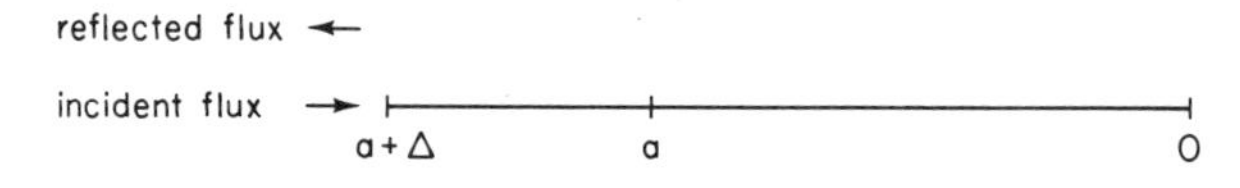

Figure 13.6

In order to obtain equations for the $r_{ij}(a)$ as functions of a, we must keep account of all of the possible ways in which interactions can contribute to a reflected flux in state i. Although at first the bookkeeping seems tedious, subsequently, we shall show that with the aid of matrix notation the multidimensional case can be treated in as direct a fashion as the one-dimensional case when the physical meaning of the matrices is appreciated. This is typical of mathematical physics when suitable notation is employed. In more complex situations, infinite-dimensional operators replace the finite-dimensional matrices.

Starting with an incident flux in state j, we can obtain a reflected flux in state i in the following fashions:

(a) Interaction in $[a + \Delta, a]$ and immediate reflection resulting from a change of direction.
(b) Interaction in $[a + \Delta, a]$ together with reflection in state i from $[a, 0]$.
(c) Passage through $[a + \Delta, a]$ without interaction, reflection from $[a, 0]$ in state i, and passage through $[a, a + \Delta]$ without interaction. (13.14.2)
(d) Passage through $[a + \Delta, a]$ without interaction, reflection from $[a, 0]$ in state $j \neq i$, and interaction in $[a, a + \Delta]$ resulting in emergence in state i.
(e) Passage through $[a, + \Delta\ a]$ without interaction, reflection from $[a, 0]$ in state k, interaction in $[a, a + \Delta]$ resulting in a state l directed toward $[a, 0]$, and then reflection in state i from $[a, 0]$.

All further interactions produce terms of order Δ^2 and can therefore be neglected at this time.

Taking account of the possibilities enumerated in (13.14.2) we may write

$$r_{ij}(a + \Delta) = b_{ij}\Delta + \sum_{k \neq i} a_{kj} r_{ik}(a)\,\Delta + (1 - a_{jj}\Delta)[r_{ij}(a)(1 - c_{ii}\Delta) + \sum_{k \neq i} r_{kj}(a) c_{ik}\Delta + \sum_{k,\ell} r_{i\ell}(a) d_{\ell k} r_{kj}(a)\Delta]. \tag{13.14.3}$$

Passing to the limit as $\Delta \to 0$, we obtain the nonlinear differential equation

$$r'_{ij}(a) = b_{ij} + \sum_{k \neq i} a_{kj} r_{ik}(a) - (a_{jj} + c_{ii}) r_{ij}(a) + \sum_{k \neq i} r_{kj}(a) c_{ik} + \sum_{k,\ell} r_{i\ell}(a) d_{\ell k} r_{kj}(a), \tag{13.14.4}$$

for $i, j = 1, 2, \ldots, N$, with the initial condition

$$r_{ij}(0) = 0. \tag{13.14.5}$$

We are not concerned with the rigorous derivation of (13.14.4) in the foregoing fashion at the present time. A rigorous derivation of (13.14.4) from the linear equations of Sec. 13.12 will be given in the next section.

Writing

$$R(a) = (r_{ij}(a)), \tag{13.14.6}$$

and letting A, B, C, D have the previous significances, we see that (13.14.4) takes the compact form

$$R' = B + RA + CR + RDR, \qquad R(0) = 0. \tag{13.14.7}$$

This is a matrix differential equation of Riccati type. Similar equations arise in modern control theory.

13.15. Equivalence of Analytic Formulations

In Sec. 13.13, we derived the expression $R(a) = Y_1(a)X_1(a)^{-1}$ for the reflection matrix. Let us show directly that R satisfies (13.14.7) in the neighborhood of $a = 0$. We have

$$R' = Y_1' X_1^{-1} - Y_1X_1^{-1}X_1' X_1^{-1}. \tag{13.15.1}$$

Referring to the equations for X_1 and Y_1 of Sec. 12 and dropping the subscripts, we have

$$\begin{aligned} R' &= (BX + CY)X^{-1} + YX^{-1}(AX + DY)X^{-1} \\ &= B + CR + RA + RDR, \end{aligned} \tag{13.15.2}$$

the desired equation.

The point of the invariant imbedding formulation is to derive an equation for R which we can use to establish its existence for $a \geqslant 0$. Without the theory of invariant imbedding to guide our steps, (13.15.2) would have little meaning.

Exercises

1. Show that (13.15.2) can be interpreted in the following fashion: B corresponds to immediate backscattering, CR corresponds to reflection followed by forward scattering, RA corresponds to forward scattering followed by transmission, and RDR corresponds to reflection, backscattering and then reflection again.
2. Show that the equation for the transmission matrix, $T(a)$, is

$T' = TA + TDR$, forward scattering, followed by transmission, plus reflection, backscattering and transmission again.

3. Derive the equation from the equations for X_1 and Y_1.

13.16. Conservation Relation

Since we have assumed a pure scattering process, we know that no particles are lost. This is equivalent to the assumption

$$M(B + A) = 0, \qquad M(C + D) = 0, \tag{13.16.1}$$

where M is the matrix

$$M = \begin{pmatrix} 1 & 1 & \dots & 1 \\ 0 & 0 & \dots & 0 \\ \vdots & & & \\ 0 & 0 & \dots & 0 \end{pmatrix}. \tag{13.16.2}$$

The effect of M is to produce column sums.

We expect then that the conservation relation

$$M(R + T) = M, \tag{13.16.3}$$

which states that total input is equal to total output, will hold for $a \geqslant 0$. Let us begin by showing that it holds for small a. Using the differential equations for R and T, we have

$$\begin{aligned}
[M(R + T)]' &= M(B + CR + RA + RDR) + M(TA + TDR) \\
&= MB + MCR + MRA + MTA + M(R + T)DR \\
&= MB + MCR + M(R + T - I)A + MA \\
&\quad + M(R + T - I)DR + MDR \\
&= M(B + A) + M(C + D)R + M(R + T - I)A \\
&\quad + M(R + T - I)DR \\
&= M(R + T - I)A + M(R + T - I)DR.
\end{aligned} \tag{13.16.4}$$

Regarding R as a known function, this is a differential equation of the form

$$Z' = ZA + ZDR, \qquad Z(0) = 0, \tag{13.16.5}$$

for the function $Z = M(R + T - I)$. One solution is clearly $Z = 0$. Uniqueness assures us that it is the only solution for small a.

Hence, within the domain of existence of R and T, we have (13.16.3). If we can establish what is clearly to be expected, namely that all the elements of R and T are nonnegative (13.16.3) will imply that R and T are uniformly bounded for $0 \leqslant a \leqslant a_0$, which means that we can continue the solution indefinitely, using the same argument repeatedly.

Exercises

1. Establish nonnegativity of R directly from the Riccati equation, using the integral equation

$$R = \int_0^t e^{C(t-t_1)} (B + RDR) e^{A(t-t_1)} dt_1 .$$

Use successive approximations and the fact that e^{At} and e^{Ct} are nonnegative matrices.)

Similarly, establish the nonnegativity of T using the equation $T' = TA + TDR$ and the result for R.

2. If $A(t)$ and $C(t)$ are variable, use the integral equation

$$R = \int_0^t X(t)X(t_1)^{-1} [B + RDR] \, Y(t)Y(t_1)^{-1} \, dt_1 .$$

3. Establish that $M(R + T) = M$ by showing that $M(Y_1(a) + I) = MX_1(a)$.

4. If A, B, C, and D are constant, show that $R(a)$ is monotone increasing as a increases. What about $T(a)$?

5. Show that $R(a)$ has an asymptotic expansion of the form

$$R(a) = R(\infty) + R_1 e^{\lambda_1 a} + \cdots, \qquad \text{where} \quad \lambda_1 < 0.$$

13.17. Internal Fluxes

Let us now turn to a discussion of $x(t)$ and $y(t)$, the solution of (13.11.1) We want to show that these internal fluxes are nonnegative vectors. To simplify the details, let us consider the homogeneous, isotropic case.

Consider Fig. 13.7. Restricting our attention to the interval $[a, t]$, we can write

$$x(t) = T(a - t)c + R(a - t)\, y(t), \tag{13.17.1}$$

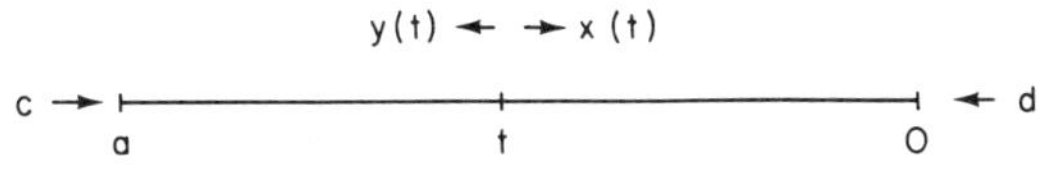

y(t) ← → x(t)

c → a t 0 ← d

Figure 13.7

while the interval $[t, 0]$ yields

$$y(t) = R(t)\, x(t) + T(t)d. \tag{13.17.2}$$

Thus

$$x(t) = T(a - t)c + R(a - t)[R(t)\, x(t) + T(t)d], \tag{13.17.3}$$

whence

$$x(t) = [I - R(a - t)\, R(t)]^{-1}\, [T(a - t)c + R(a - t)\, T(t)d]. \tag{13.17.4}$$

The fact that $MR < M$ means that

$$M(R(a - t)\, R(t)) = (MR(a - t))\, R(t) < MR(t) < M. \tag{13.17.5}$$

Hence the series

$$(r - R(a - t)\, R(t))^{-1} = r + R(a - t)\, R(t) + (R(a - t)\, R(t))^2 + \cdots \tag{13.17.6}$$

converges, showing that $x(t)$ is positive for $0 < t < a$.

13.18. Asymptotic Behavior

Let us next turn to some questions of asymptotic behavior.
Consider the scalar equation

$$u' = -u + g(u), \qquad u(0) = c, \tag{13.18.1}$$

where $|g(u)| = o(|u|)$ as $u \to 0$ and $|c| << 1$. We proved in Chapter 4 that under these conditions the solution exists for all $t \geqslant 0$, and furthermore that

$$u \sim e^{-t}\varphi(c) \tag{13.18.2}$$

as $t \to \infty$. In addition if $g(u)$ is analytic in $|u|$ for $|u|$ small, then $\varphi(c)$ is analytic in c for small $|c|$. However, we provided no convenient effective technique for calculating $\varphi(c)$.

Let us now employ invariant imbedding to obtain a differential equation for $\varphi(c)$ which permits its ready calculation. We begin by writing

$$\varphi(c) = \lim_{t\to\infty} u(t)e^t. \tag{13.18.3}$$

This function is defined for $| c | << 1$. Then

$$\varphi(c) = \lim_{t\to\infty}[u(t + \Delta)e^{t+\Delta}] = e^{\Delta} \lim_{t\to\infty} u(t + \Delta)e^{t}. \tag{13.18.4}$$

Using the definition of $\varphi(c)$ in (13.18.3) and the differential equation in (13.18.1) this may be written

$$\varphi(c) = e^{\Delta}\varphi(c + (-c + g(c))\Delta) + O(\Delta^2). \tag{13.18.5}$$

Hence, expanding in powers of Δ,

$$\begin{aligned}\varphi(c) &= (1 + \Delta)[\varphi(c) - (c + g(c))\varphi'(c)\Delta] + O(\Delta^2) \\ &= \varphi(c) + \Delta\varphi(c) - (c + g(c))\varphi'(c)\Delta + O(\Delta^2).\end{aligned} \tag{13.18.6}$$

Thus, in the limit as $\Delta \to 0$, we obtain the differential equation

$$\varphi'(c) = \frac{\varphi(c)}{c + g(c)}. \tag{13.18.7}$$

Returning to (13.18.1), we know that

$$u = e^{-t}c + O(c^2). \tag{13.18.8}$$

Hence, a suitable normalization for (13.18.7) is

$$\varphi(c) = c + O(c^2). \tag{13.18.9}$$

Setting $\varphi(c) = c + \sum_{n=2}^{\infty} a_n c^n$, we can readily use (13.18.7) to obtain a_2 and generally a recurrence relation for a_n in terms of $a_2, \ldots, a_{n-1}$, and thus determine the coefficients as far out as desired.

Exercises

1. Calculate the first three terms of $\varphi(c)$ for the case where $g(u) = u^n$, with n an integer greater than 1.
2. Obtain the result from the explicit solution of (13.18.1) in integral form.
3. What is the radius of convergence of $\varphi(c)$?
4. What is the interval of stability for (13.18.1)? Is it equal to the radius of convergence of $\phi(c)$?

13.19. Multidimensional Case

Exactly the same method can be applied to the vector case where

$$\frac{dx}{dt} = Ax + g(x), \qquad x(0) = c, \tag{13.19.1}$$

and we suppose that the real parts of the characteristic roots of A are negative, i.e., that A is a stability matrix, that $\| c \| < < 1$, and that $g(x)$ is analytic in the components of x in some neighborhood of the origin, lacking constant and first degree terms.

Consider the case where the root with largest real part is real. Call this root λ. Then we define the function $\varphi(c)$ by means of the relation

$$\varphi(c) = \lim_{t\to\infty} e^{-\lambda t}\, x(t), \tag{13.19.2}$$

and proceed as before to obtain the equations

$$\begin{aligned} \varphi(c) &= e^{-\lambda\Delta}\, \varphi(c + [Ac + g(c)]\Delta) + O(\Delta^2) \\ &= \varphi(c) - \lambda\varphi(c)\Delta + J(\varphi)[Ac + g(c)]\Delta + O(\Delta^2) \end{aligned} \tag{13.19.3}$$

(here $J(\varphi)$ is the Jacobian matrix), and thus, in the limit,

$$\lambda\varphi(c) = J(\varphi)[Ac + g(c)]. \tag{13.19.4}$$

The normalization for φ is furnished by the asymptotic relation obtained from (13.19.1),

$$x = e^{At}c + o(\| c \|)e^{\lambda t}, \tag{13.19.5}$$

as $t \to \infty$. Thus

$$\varphi(c) = Bc + o(\| c \|), \tag{13.19.6}$$

where B is a known matrix.

Exercises

1. Consider the equations

$$\begin{aligned} u' &= -u - v + u^2, & u(0) &= c_1, \\ v' &= -2v + uv, & v(0) &= c_2. \end{aligned}$$

Determine the first and second degree terms in $\varphi(c) \equiv \varphi(c_1, c_2)$.

2. To this order of approximation, determine the curve in the (c_1, c_2)-plane, (the phase-plane), with the property that for points on this curve the asymptotic form of the solution is $u \sim a(c_1, c_2)e^{-2t}$, $v \sim b(c_1, c_2)e^{-2t}$, and determine $a(c_1, c_2)$ and $b(c_1, c_2)$.

3. Obtain the asymptotic behavior when the equations are

$$u' = -u - v + u^2, \qquad u(0) = c_1,$$
$$v' = -v + uv, \qquad v(0) = c_2.$$

4. Consider the nonlinear differential equation

$$u' + u + u^2 = c_2 e^{-at}, \qquad u(0) = c_1, \qquad a > 0, \quad |c_1|, |c_2| \neq 1.$$

Determine $u(c_1, c_2, t)$ as a power series in c_1 and c_2 by considering the associated system

$$u' + u + u^2 = v, \quad u(0) = c_1, \qquad v' + av = 0, \quad v(0) = c_2.$$

5. Consider similarly the equation

$$u' + u + u^2 = c_1 e^{-a_1 t} + \cdots + c_M e^{-a_M t}.$$

13.20. Asymptotic Phase

The same methods can be applied to study the asymptotic behavior of the solutions of

$$u'' + (1 + f(t))u = 0, \qquad u(0) = c_1, \quad u'(0) = c_2, \qquad (13.20.1)$$

under various assumptions concerning $f(t)$. In Chapter 1, we showed that the condition

$$\int_0^\infty |f| \, dt < \infty$$

ensured that the solutions of (13.20.1) are bounded as $t \to \infty$ for any particular set of values c_1 and c_2, and furthermore that we have the simple asymptotic behavior

$$u \sim (b_1 \cos t + b_2 \sin t) \qquad (13.20.2)$$

as $t \to \infty$ for some constants b_1 and b_2. However, we skirted the question of an effective determination of b_1 and b_2 in terms of c_1 and c_2.

Let us sketch the principal idea of a procedure based on invariant imbedding and refer to original sources for further details. We use the exponential form of (13.20.2) which simplifies the algebra to some extent. Write

$$u(t) \sim e^{i(t-a)} g(c_1, c_2, a) + e^{-i(t-a)} h(c_1, c_2, a),$$
$$u'(t) \sim ie^{i(t-a)} g(c_1, c_2, a) - ie^{-i(t-a)} h(c_1, c_2, a), \tag{13.20.3}$$

where we use a generic point a as the origin in time,

$$u'' + (1 + f(t))u = 0, \qquad u(a) = c_1, \quad u'(a) = c_2. \tag{13.20.4}$$

The inhomogeneity of the process in time forces us to use this variable. The fact, however, that we know that $g(c_1, c_2, a)$ and $h(c_1, c_2, a)$ are linear in c_1 and c_2 will yield a significant reduction in complexity.

Let us begin by dealing directly with (13.20.3) We have

$$\lim_{t\to\infty}\left[\frac{u(t) + iu'(t)}{2}\right]e^{i(t-a)} = h(c_1, c_2, a),$$
$$\lim_{t\to\infty}\left[\frac{u(t) - iu'(t)}{2}\right]e^{-i(t-a)} = g(c_1, c_2, a). \tag{13.20.5}$$

Hence, proceeding as before,

$$h(c_1, c_2, a) = e^{2i\Delta}h(c_1 + c_2\Delta, c_2 - \Delta(1 + f(a))c_1, a + \Delta) + O(\Delta^2)$$
$$g(c_1, c_2, a) = e^{-2i\Delta}g(c_1 + c_2\Delta, c_2 - \Delta(1 + f(a))c_1, a + \Delta) + O(\Delta^2). \tag{13.20.6}$$

Expanding and passing to the limit, we obtain the partial differential equations

$$0 = 2ih + c_2\frac{\partial h}{\partial c_1} - (1 + f(a))c_1\frac{\partial h}{\partial c_2} + \frac{\partial h}{\partial a},$$
$$0 = -2ig + c_2\frac{\partial g}{\partial c_1} - (1 + f(a))c_1\frac{\partial g}{\partial c_2} + \frac{\partial g}{\partial a}. \tag{13.20.7}$$

Setting

$$h(c_1, c_2, a) = h_1(a)c_1 + h_2(a)c_2,$$
$$g(c_1, c_2, a) = g_1(a)c_1 + g_2(a)c_2, \tag{13.20.8}$$

we obtain ordinary differential equations in a for h_1, h_2, g_1, g_2.

The boundary conditions for large a are obtained from the asymptotic expansion of the solutions of (13.20.1) for large t. To obtain the values for $a = 0$, we integrate backwards.

Exercises

1. Write out the differential equations for h_1, h_2, g_1, g_2.
2. Obtain the boundary conditions for large a.
3. Discuss the computational feasibility of this approach to the determination of $g_1(0)$, $g_2(0)$, $h_1(0)$, $h_2(0)$.
4. Obtain the corresponding results for the matrix case $X'' + AX = 0$, $X(0) = c_1$, $X'(0) = c_2$, where A is positive definite.
5. Write $u'' + (1 + f(t))u = 0$, $u(0) = \cos\theta$, $u'(0) = \sin\theta$, and obtain differential equations for the functions $g(\theta)$, $h(\theta)$, and $\varphi(\theta)$, where $u(t) \sim e^{it}g(\theta) + e^{-it}h(\theta) = \cos(t + \varphi(\theta))$. See

   G. M. Wing, "Invariant Imbedding and the Asymptotic Behavior of Solutions to Initial-value Problems," *J. Math. Anal. Appl.*, Vol. 9, 1964, pp. 85–98.

   W. A. Beyer, "Asymptotic Phase and Amplitude for a Modified Coulomb Potential in Scattering Theory: An Application of Invariant Imbedding," *J. Math. Anal. Appl.*, Vol. 13, 1966, pp. 348–360.
6. Consider the equation $u'' - (1 + ze^{-\lambda t})u = 0$, $u(0) = c_1$, $u'(0) = c_2$, and define $f_1(z)$ and $f_2(z)$ by $\lim_{t\to\infty} ue^{-t} = c_1 f_1(z) + c_2 f_2(z)$. Obtain linear differential equations satisfied by f_1 and f_2.
7. Use these equations to obtain power series expansions for f_1 and f_2 as functions of z.
8. Similarly consider the vector matrix case $x' - (A + e^{-Bt}Z)x = 0$. See

   R. Bellman, "On Asymptotic Behavior of Solutions of Second-Order Differential Equations," *Quart. Appl. Math.*, Vol. 20, 1963, pp. 385–387.
9. Consider the equation $u'' + (1 + \lambda g(t))u = 0$, where $\int^\infty |g(t)|\, dt < \infty$ and write $u \sim r(\lambda)\sin(t + \phi(\lambda))$ as $t \to \infty$. Consider λ to be a complex variable and r, ϕ as functions of λ. What are the analytic properties of $r(\lambda)$ and $\theta(\lambda)$? Where are the singularities nearest the origin? (Kemp–Levinson.)

13.21. Wave Propagation

The wave equation

$$k(x)w_{tt} = w_{xx} \tag{13.21.1}$$

reduces to an ordinary differential equation

$$u'' + k(x)u = 0, \tag{13.21.2}$$

if we set $w(x, t) = u(x)e^{it}$. If $k(x)$ is constant, the solution is readily obtained; if $k(x)$ is variable, we need some convenient analytic approximation. The most famous of these, the WKB approximation (more properly the Liouville approximation), was given in Chapter 1, Volume I. Let us now indicate a physical interpretation of this approximation.

If there are two semi-infinite homogeneous regions abutting (see Fig. 13.8), which is to say

$$\begin{aligned} k(x) &= k_1^2, \qquad -\infty < x \leqslant a, \\ &= k_2^2, \qquad a < x < \infty, \end{aligned} \tag{13.21.3}$$

- ∞ a + ∞

Figure 13.8

we can readily determine the effect of an incoming wave, say

$$u = e^{ik_1x}, \qquad -\infty < x < a. \tag{13.21.4}$$

We have a solution of the form

$$\begin{aligned} u &= e^{ik_1x} + re^{-ik_1x}, \qquad -\infty < x < a, \\ u &= te^{ik_2x}, \qquad a < x < \infty, \end{aligned} \tag{13.21.5}$$

with the quantities r and t determined by the postulated continuity of u and u' at $x = a$.

If we take the interface to be at $x = 0$, the continuity conditions yield the results

$$\begin{aligned} u_1(0) &= 1 + r = t = u_2(0), \\ u_1'(0) &= ik_1(1 - r) = ik_2t = u_2'(0). \end{aligned} \tag{13.21.6}$$

Hence

$$r = \frac{k_1 - k_2}{k_2 + k_1}, \qquad t = \frac{2k_1}{k_2 + k_1}. \tag{13.21.7}$$

When the interface is at $x = a$ (Fig. 13.8), we write the solutions in the form

$$u_1 = e^{ik_1(x-a)} + c_1e^{-ik_1(x-a)}, \qquad u_2 = c_2e^{ik_2(x-a)}. \tag{13.21.8}$$

Consider next the case where $k(x)$ varies as a function of x for $0 < x < \infty$ and we begin once again with the incoming wave of (13.21.4). To obtain an approximate result, we can divide the interval $[a, y]$ into a finite number of parts and suppose that $k(x)$ is constant in each of these. Piecing together the solutions found in this way, we can obtain an approximate solution for $u(x)$ at $x = y$. We thus obtain the WKB approximation in the following fashion (see Fig. 13.9). Suppose that

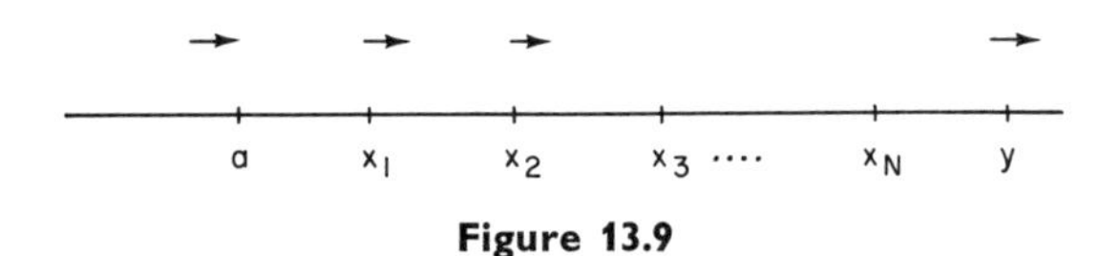

Figure 13.9

$k(x)$ varies slowly with x so that at each interface the reflected intensity is small. As a first approximation, assume that the intensity at y is the result only of transmissions across the interfaces with no attention paid to any reflections and subsequent transmissions.

Then the amplitude of the direct transmitted wave just to the left of $x = x_N$ is given by

$$a_N = \left(\frac{2k_0}{k_0 + k_1}\right)\left(\frac{2k_1}{k_1 + k_2}\right) \cdots \left(\frac{2k_{N-1}}{k_{N-1} + k_N}\right) \exp\left(i \sum_{i=1}^{N} k_j \,\Delta x_j\right). \tag{13.21.9}$$

Let $\Delta x_i = \Delta$ and $N\Delta \to x$ as $\Delta \to 0$, then a_N approaches the quantity

$$u_0(x) = \left(\frac{k_0}{k(x)}\right)^{1/2} \exp\left(i \int_0^x k(s)\, ds\right). \tag{13.21.10}$$

We recognize this as the famous WKB approximation.

13.22. The Bremmer Series

The importance of the foregoing interpretation lies in the fact that this simple physical model of successive reflections and transmissions can be used to obtain further approximations, and actually a solution, in a very simple fashion. Furthermore, it generalizes to higher dimensions and other processes.

Following Bremmer, we obtain these approximations by combining the effects of successive reflections. The contribution at x due to these waves which are the result of one reflection is easily seen to be

$$u_1(x) = -\frac{1}{2k(x)^{1/2}} \int_x^\infty \frac{k'(s)}{k(s)^{1/2}}\, u_0(s) \exp\left\{i \int_x^s k(t)\, dt\right\} ds. \tag{13.22.1}$$

Similarly, if we define $u_n(x)$ as the contribution due to those waves reflected n times, we see that

$$u_{2n}(x) = \frac{1}{2(k(x))^{1/2}} \int_0^x \frac{k'(s)}{k(s)^{1/2}} u_{2n-1}(s) \exp\left\{i \int_s^x k(t)\, dt\right\} ds,$$
$$u_{2n+1}(x) = -\frac{1}{2k(x)^{1/2}} \int_x^\infty \frac{k'(s)}{k(s)^{1/2}} u_{2n}(s) \exp\left\{i \int_x^s k(t)\, dt\right\} ds. \qquad (13.22.2)$$

It follows that a solution of

$$u'' + k^2(x)u = 0 \qquad (13.22.3)$$

should be obtained by taking account of the total contribution of all reflections.

We call the expression

$$u(x) = \sum_{i=0}^{\infty} u_i(x) \qquad (13.22.4)$$

formed in this way, the *Bremmer series.*

Exercises

1. Establish (13.22.2).
2. Show that if (a) $|\, k(x)| \geqslant a > \infty$, (b) $\int_0^\infty |\, k'(x)|\, dx < 2a$, then the Bremmer series converges to a solution of (13.22.3). (We do not pursue this in detail since the result can be substantially improved: see the exercises at the end of the chapter.) In particular, however, $\int^\infty |\, k'(x)|\, dx < \infty$ is not sufficient to ensure convergence. See, for the above

   R. Bellman, and R. Kalaba, "Functional Equations, Wave Propagation and Invariant Imbedding," *J. Math. Mech.*, Vol. 8, 1959, pp. 683–702.

13.23. Localization Principles

The study of particle motion by way of functional equation techniques depends upon a very simple but important property of this process, of such simplicity its basic nature is usually overlooked. Consider, to illustrate this point, a particle traveling along the line segment [a, b], suffering collisions which may cause it to reverse direction, change energy and so on. In deriving equations describing its position as a function of time, we use the fact that its behavior in $[a + \Delta, a]$ depends only upon the particular physical properties of this segment. This is a *localization principle* which we take for granted in studying particle processes.

Another way of stating this principle is to postulate that the particle constantly behaves as if the entire interval ahead of it were homogeneous, possessing the properties of the infinitesimal interval directly in front of it. This is not uniformly true as the case of the motion of a charged particle through a plasma shows. Like all useful principles, it is sometimes true and sometimes not.

13.24. Localization for Wave Motion

The processes leading to the Bremmer series permit us to state a corresponding principle for wave motion.

Principle of Localization. *A plane wave traveling through an inhomogeneous medium proceeds as if there were an instantaneous reflection and transmission at each interface of a stratum* $[y, y + \Delta]$. *These reflections and transmissions occur as if the stratum* $[y, y + \Delta]$ *were actually a semi-infinite homogeneous medium,* $[y, \infty]$, *with wave number* $k(y)$ *for* $x \geqslant y$.

The convergence of the Bremmer series would permit an easy establishment of this localization principle. Unfortunately, as indicated above, the Bremmer series does not always converge. In order then to establish the result in general, we use a different method, one compounded of both the classical approach and that of invariant imbedding.

13.25. The Classical Solution to Wave Propagation through an Inhomogeneous Medium

We require some standard results concerning reflected and transmitted waves. In the interval $[-\infty, z]$ we have the equation

$$u'' + k_1^2 u = 0 \tag{13.25.1}$$

with a normalized solution of the form

$$u = e^{ik_1(x-z)} + c_1 e^{-ik_1(x-z)}. \tag{13.25.2}$$

We think of the term $e^{ik_1(x-z)}$ as representing an incoming wave from

$-\infty$, while $c_1 e^{-ik_1(x-z)}$ represents the reflected wave returning to $-\infty$ due to the change in the medium at $x = z$ (Fig. 13.10). The coefficient $c_1 = c_1(z)$ is called the *reflection coefficient*.

k_1^2 $k^2(x)$ k_2^2

$-\infty$ z b $+\infty$

Figure 13.10

In $[z, b]$ the equation assumes the form

$$u'' + k^2(x)u = 0 \tag{13.25.3}$$

with a solution of the form

$$u = c_2 u_1(x) + c_3 u_2(x) \tag{13.25.4}$$

where $u_1(x)$ and $u_2(x)$ are particular linearly independent solutions, determined for the sake of convenience by the conditions

$$u_1(b) = 1, \quad u_1'(b) = 0; \qquad u_2(b) = 0, \quad u_2'(b) = 1. \tag{13.25.5}$$

In other words, u_1 and u_2 are principal solutions in $[b, z]$.

In $[b, \infty]$ we have the equation

$$u'' + k_2^2 u = 0 \tag{13.25.6}$$

and we postulate the solution

$$u = c_4 e^{ik_2(x-b)} \tag{13.25.7}$$

a wave going to $+\infty$ where c_4 is a coefficient of *transmission*. There is no wave coming from $+\infty$.

To determine the four coefficients c_1, c_2, c_3, and c_4, we further postulate continuity of the solution and its derivative at the interfaces, $x = z$, $x = b$. We thus obtain the conditions

$$c_2 = c_4, \qquad c_3 = ik_2 c_4, \tag{13.25.8}$$

at $x = b$, and

$$1 + c_1 = c_4(u_1(z) + ik_2 u_2(z)), \quad ik_1(1 - c_1) = c_4(u_1'(z) + ik_2 u_2'(z)) \tag{13.25.9}$$

at $x = z$.

This yields

$$c_1 = -\frac{(\phi'(z)/\phi(z) - ik_1)}{(\phi'(z)/\phi(z) + ik_1)}, \tag{13.25.10}$$

with c_2, c_3, and c_4 also determined where $\phi(z) = u_1(z) + ik_2u_2(z)$.

13.26. Riccati Equation

Since $\phi(x)$ satisfies the linear equation (13.25.3), the function $w(z) = \phi'(z)/\phi(z)$ satisfies the Riccati equation

$$w' + w^2 + k^2(z) = 0. \tag{13.26.1}$$

Let us next assume that $k(x)$ is differentiable and that $k(z) = k_1$. From (13.25.10),

$$w = ik_1 \frac{1 - c_1}{1 + c_1} = ik(z) \left(\frac{1 - c_1(z)}{1 + c_1(z)}\right). \tag{13.26.2}$$

Thus, $c_1(z)$ satisfies the Riccati equation

$$c_1'(z) = -2ik(z)c_1(z) + \frac{k'(z)}{2k(z)}(1 - c_1^2(z)). \tag{13.26.3}$$

As we shall see, this is precisely the result obtained using a combination of invariant imbedding and the principle of localization.

13.27. Invariant Imbedding

The problem we have been considering is that of determining the reflected and transmitted waves associated with a plane parallel slab, homogeneous in the y- and z-directions, but inhomogeneous in the x-direction. We consider only normally incident plane waves. To perform the imbedding, we keep the right interface fixed at $x = b$ and allow the left interface to be variable, $x = z$, where $-\infty < z < b$. The intensities of the transmitted and reflected waves will thus be functions of z (Fig. 13.11).

The medium to the left of $x = z$ will be taken to be homogeneous with wave number $k(z)$, or more precisely $k(z - 0)$, since there could be a step discontinuity at $x = z$. The medium to the right of $x = b$ will be taken to be homogeneous with wave number $k(b)$, again $k(b + 0)$. We write $k_1 = k(b)$.

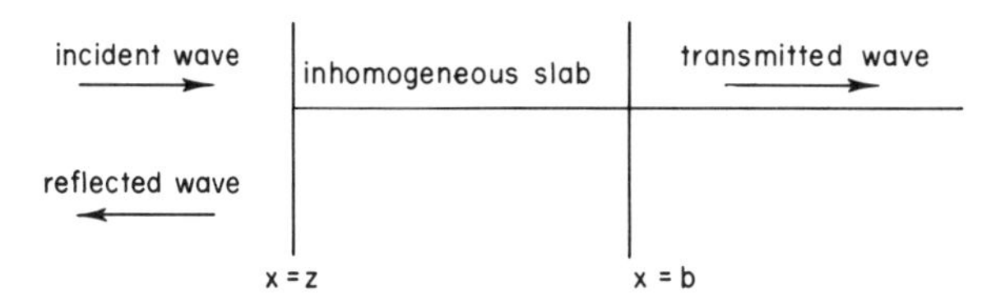

Figure 13.11

We now wish to show that the use of localization ideas can be validated using invariant imbedding. The reflected wave will have the form $u(z)e^{-ik(z)(x-z)}$ where $u(z)$ is the function we wish to determine.

13.28. Functional Equations

Consider the schematic in Fig. 13.12. In the infinitesimal interval $[z, z + \Delta]$, we take the wave number $k(x)$ to have the constant value $k(z + \Delta)$. Taking account of zeroth and first order terms in Δ, the total reflected wave from $[z, b]$ will be a sum of three parts:

(a) The immediate reflected wave from $[z, z + \Delta]$.

(b) The wave that arises from transmission through $[z, z + \Delta]$, reflection from the interface $x = z + \Delta$, and transmission through the interface $x = z$.

(c) The wave that results from transmission through $x = z$, reflection at $x = z + \Delta$, reflection again at $x = z$, and finally transmission through $x = z$.

Schematically, the cases are shown in Fig. 13.13.

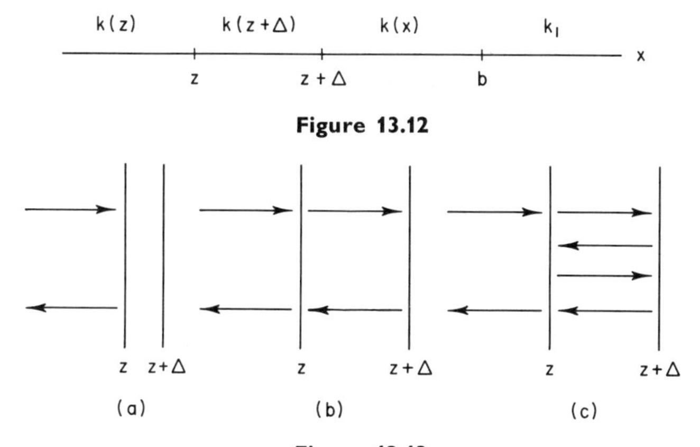

Figure 13.13

The intensities of the reflected and transmitted waves are obtained in terms of the function $u(z)$ defined above and the principle of localization. An analytic transliteration is the equation

$$\begin{aligned}
u(z) = {} & \frac{k(z-0) - k(z+\Delta-0)}{k(z-0) + k(z+\Delta-0)} \\
& + \frac{2k(z-0)}{k(z-0)+k(z+\Delta-0)} u(z+\Delta) e^{2ik(z+\Delta-0)\Delta} \\
& \cdot \frac{2k(z+\Delta-0)}{k(z-0)+k(z+\Delta-0)} \\
& + \frac{2k(z-0)}{k(z-0)+k(z+\Delta-0)} e^{ik(z+\Delta-0)\Delta} u(z+\Delta) e^{ik(z+\Delta-0)\Delta} \\
& \cdot \frac{k(z+\Delta-0)-k(z-0)}{k(z-0)+k(z+\Delta-0)} e^{ik(z+\Delta-0)\Delta} u(z+\Delta) e^{ik(z+\Delta-0)\Delta} \\
& \cdot \frac{2k(z+\Delta-0)}{k(z+\Delta-0)+k(z-0)} + o(\Delta).
\end{aligned} \tag{13.28.1}$$

Passing to the limit as $\Delta \to 0$, we obtain the Riccati equation

$$u'(z) = \frac{k'(z)}{2k} - 2iku - \frac{k'}{2k} u^2, \tag{13.28.2}$$

precisely the equation derived in Sec. 13.26 using the conventional approach.

Exercise

1. Let the transmitted wave have the form $u(x) = v(b)e^{ik_1(x-b)}$, $x > b$. Show that

$$v' = \left(\frac{k'}{2k} - u\frac{k'}{k} - ik\right)v$$

using the conventional approach and invariant imbedding.

13.29. Random Walk

Let us next apply invariant imbedding concepts to the discussion of some equations connected with random walk. Consider the following stochastic process. A point particle can occupy any of the lattice points in the finite interval $[a, b]$ where a and b are themselves lattice points (Fig. 13.14).

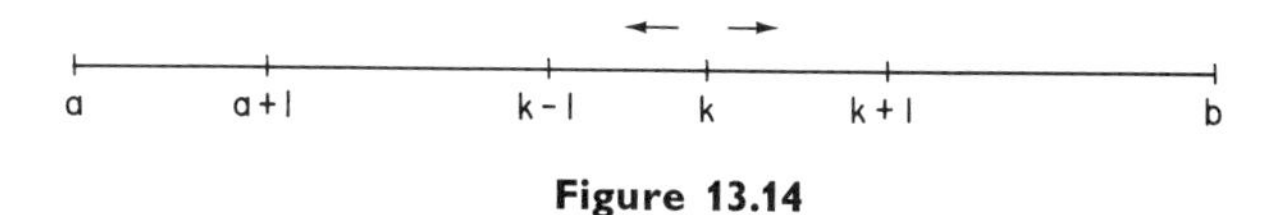

Figure 13.14

At discrete times, $t = 0, 1, 2, \ldots$, the particle moves either to the right or to the left a distance of one unit. Let

$$\begin{aligned} p(k) &= \text{the probability at } k \text{ of moving one unit to the left at } k, \\ q(k) &= \text{the probability at } k \text{ of moving one unit to the right at } k, \end{aligned} \tag{13.29.1}$$

where $p(k) + q(k) = 1$, $0 < p(k) < 1$, $a < k < b$.

We suppose that the process terminates if the particle lands at either a or b, hence the name *absorbing barriers*. It is not difficult to show that with probability one the particle will hit an absorbing barrier at some finite time. The problem we pose is that of determining the probability that a particle starting at k will land at a before b.

13.30. Classical Formulation

Let us begin with the classical imbedding. Let

$$u(k) = \text{the probability that a particle starting at } k \text{ lands at } a \text{ before landing at } b. \tag{13.30.1}$$

The function is defined for $k = a + 1, a + 2, \ldots, b - 1$, and we set

$$u(a) = 1, \qquad u(b) = 0. \tag{13.30.2}$$

Taking account of the possible outcomes of the first event, we readily obtain the equation

$$u(k) = p(k)u(k - 1) + q(k)u(k + 1), \qquad a < k < b. \tag{13.30.3}$$

Exercises

1. Show that (13.30.3) possesses a unique solution.

2. Consider the possibility that the particle can move as before and, in addition, remain stationary with probability $r(k)$. Show that now $u(k) = p(k)u(k-1) + q(k)u(k+1) + r(k)u(k)$, and thus that this process is equivalent to the original with $p(k)$ replaced by $p(k)/(1 - r(k))$ and $q(k)$ by $q(k)/(1 - r(k))$.

13.31. Invariant Imbedding

Let us now make the observation that $u(k)$ depends upon both a and b. Let us keep b fixed and regard a as variable. Thus, we are asking for the probability of escape to the left as a function both of the position of the particle and the length of the rod. Let us then introduce the function

$$u(a, k) = \text{the probability that a particle at } k \text{ lands at } a \text{ before landing at } b. \tag{13.31.1}$$

The function is defined for $-\infty < a < k < b$. We set

$$u(a, a) = 1, \qquad u(a, b) = 0. \tag{13.31.2}$$

To obtain an equation for $u(a, k)$ we use the fact that a particle reaching a starting from k must necessarily reach $a + 1$ at some intermediate time. This is a consequence of the fact that a particle can move only one unit in either direction (Fig. 13.15). From this follows the

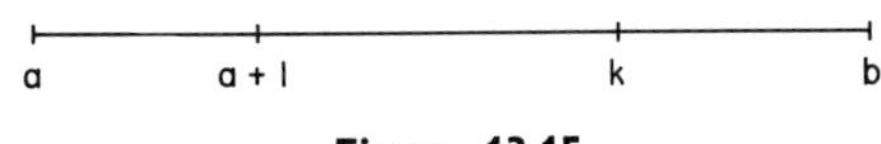

Figure 13.15

analytic relation

$$u(a, k) = u(a + 1, k)\, u(a, a + 1). \tag{13.31.3}$$

Iterating this relation, we see that

$$u(a, k) = u(a, a + 1)\, u(a + 1, a + 2) \cdots u(k - 1, k). \tag{13.31.4}$$

Thus the determination of the function of two variables $u(a, t)$, and the solution of the original problem, hinges upon our ability to determine the function $u(a, a + 1)$, $a < b$.

13.32. Determination of $u(a, a + 1)$

To obtain a relation for $u(a, a + 1)$, we return to the original process. This yields

$$u(a, a + 1) = p(a + 1) + q(a + 1)\, u(a, a + 2). \tag{13.32.1}$$

However, we can also write, by virtue of the foregoing,

$$u(a, a+2) = u(a, a+1)\, u(a+1, a+2). \tag{13.32.2}$$

Combining (13.32.1) and (13.32.2), we obtain the relation

$$u(a, a+1) = \frac{p(a+1)}{1 - q(a+1)\, u(a+1, a+2)}, \tag{13.32.3}$$

which upon iteration yields

$$\cfrac{p(a+1)}{1 - \cfrac{q(a+1)\, p(a+2)}{1 - \cfrac{q(a+2)\, p(a+3)}{1 - \cdots}}}$$

The continued fraction terminates after a finite number of steps since $u(b-1, b) = 0$. This continued fraction is a discrete version of the Riccati equation.

Exercises

1. What happens if there is a probability, p_b, that the barrier at b is not totally absorbing but bounces the particle back to $b - 1$?
2. What happens if there is a probability a_k that the particle is returned to k when b is reached?
3. Obtain the approximations

$$u(a, a+1) > \frac{p(a+1)}{1 - q(a+1)\, p(a+2)} > \cfrac{p(a+1)}{1 - \cfrac{q(a+1)\, p(a+2)}{1 - q(a+2)\, p(a+3)}}.$$

4. What are the physical meanings of these approximations?
5. Can we obtain upper bounds for $u(a, k)$ by interchanging the roles of a and b and using lower bounds of the foregoing type.
6. Introduce the function $w(a, k)$ defined as the expected time to reach a starting at k in the case where a is reached before b. Show that

$$w(a, a+1) = \frac{1}{p(a+1)} + \frac{q(a+1)}{p(a+2)\, p(a+1)} + \frac{q(a+1)\, q(a+2)}{p(a+3)\, p(a+2)\, p(a+1)} + \cdots.$$

The series terminates if b is finite.

7. Obtain an analytic solution to the conventional equation if $p(a+1) = p$, $q(a+1) = q$.

8. What is the physical significance of the result for $b = \infty$ in each of the cases $p/q > 1$, $p/q < 1$, $p = q$?

9. Consider the case where $q(a) = q^a$, $0 < q < 1$.

10. Let $z(a, k)$ be the random variable equal to the time spent going from k to a before hitting b in the case where a is reached first. Show that $z(a, k) = z(a+1, k) + z(a, a+1)$. Let $y(a, k) = y(a, k, s) = E(e^{isz})$. Then

$$y(a, a+1, s) = \frac{p(a+1)e^{is}}{[1 - q(a+1)e^{is}\, g(a+1, a+2)s)]}.$$

11. Obtain recurrence relations for $E(z^2)$.

12. Set $p(k) = \frac{1}{2} - p_1(k)\Delta$, $q(k) = \frac{1}{2} + p_1(k)\Delta$ and consider the limiting case $\Delta \to 0$.

Miscellaneous Exercises

1. Consider the system $u' = v(1 - bu)$, $v' = u(bv - 1)$, $0 \leqslant z \leqslant x$, with $u(0) = 0$, $v(x) = y$, where $b > 0$. Write $u = u_0 + bu_1 + b^2u_2 + \cdots$, $v = v_0 + bv_1 + b^2v_2 + \cdots$, and show that $u_0(z) = y \sin z/\cos x$, $v_0(z) = y \cos z/\cos x$, $0 \leqslant z < x < \pi/2$. This apparently shows that there is a critical length for small b. Show, nonetheless, that the solution of the nonlinear system exists for all $x > 0$ for any $b > 0$. See

   R. Bellman, R. Kalaba, and G. M. Wing, "Invariant Imbedding and Neutron Transport Theory—III: Neutron-neutron Collision Processes," *J. Math. Mech.*, Vol. 8, 1959, pp. 249–261.

   The point of the example is twofold. In the first place, it shows that a small amount of collision interaction can destroy the phenomenon of criticality. Secondly, it shows the danger of using perturbation techniques without careful analysis. Observe that the situation is similar to that encountered in the Burgers equation, $u_t + uu_x = bu_{xx}$, which does not exhibit a shock for $b > 0$, a viscosity effect, although the equation $u_t + uu_x = 0$ does.)

2. Let u, v, be positive integers and $f(u, v)$ be the number of steps in the Euclidean algorithm. Then $f(u, v) = 1 + f(v, u - v[u/v])$.

3. Consider the problem of maximizing the function $(x, Bx) + (y, By) + 2(x, Ay) - 2(u, x) - 2(v, y)$ over x and y where B is nonnegative definite and A is merely a real matrix. Show that the variational equations are $Bx + Ay = u$, $A'x + By = v$, and thus that the equation $Ay = u$, not necessarily of variational origin has thus been imbedded within a family of variational equations.

   R. Bellman and S. Lehman, "Functional Equations in the Theory of Dynamic Programming—IX: Variational Analysis, Analytic Continuation, and Imbedding of Operators," *Proc. Nat. Acad. Sci. U.S.A.*, Vol. 44, 1958, pp. 905–907.

4. Show that the convergence of the Bremmer series can be associated with the convergence of the Liouville–Neumann series of a linear integral equation (Atkinson).

Bibliography and Comments

§13.1. The origins of invariant imbedding lie in several different sources. Let us cite dynamic programming as already indicated, the "point of regeneration" technique of the theory of branching processes,

R. Bellman and T. E. Harris, "On the Theory of Age-dependent Stochastic Branching Processes," *Proc. Nat. Acad. Sci. USA*, Vol. 34, 1948, pp. 601–604.

T. E. Harris, *The Theory of Branching Processes*, Springer–Verlag, Berlin and New York, 1963.

L. Janossy, *Cosmic Rays*, Oxford Univ. Press, London and New York, 1950.

the "principles of invariance" of Ambarzumian and Chandrasekhar,

V. A. Ambarzumian, "Diffuse Reflection of Light by a Foggy Medium," *C. R. Acad. Sci., U.R.S.S.*, Vol. 38, No. 8, 1943, pp. 229–232.

V. A. Ambarzumian, *et al.*, *Teoreticheskaya Astrofizika*, Moscow, 1952.

S. Chandrasekhar, *Radiative Transfer*, Oxford Univ. Press, London and New York, 1950,

and the work by Redheffer in potential theory and probability theory; see

R. Redheffer, "Novel Uses of Functional Equations," *J. Rat. Mech. Anal.*, Vol. 3, 1954, pp. 271–279.

For illustrations of the applications of invariant imbedding to mathematical physics, see the books

R. Bellman, R. Kalaba, and M. Prestrud, *Invariant Imbedding and Radiative Transfer in Slabs of Finite Thickness*, American Elsevier, New York, 1963.

R. Bellman, H. Kagiwada, R. Kalaba, and M. Prestrud, *Invariant Imbedding and Time-dependent Processes*, American Elsevier, New York, 1964.

G. M. Wing, *An Introduction to Transport Theory*, Wiley, New York, 1962.

R. Preisendorfer, *Radiative Transfer on Discrete Spaces*, Pergamon, Oxford, 1965.

R. Bellman, *Vistas of Modern Mathematics: Dynamic Programming, Invariant Imbedding, and the Mathematical Biosciences*, Univ. of Kentucky Press, Lexington, Kentucky, 1968.

R. Bellman, *Invariant Imbedding: Semigroups in Time, Space and Structure, Conference on Applications of Numerical Analysis, Lecture Notes in Mathematics*, 228, Springer–Verlag, Berlin.

§13.4. The first statement of this is given in

R. Bellman, and R. Kalaba, "On the Principle of Invariant Imbedding and Propagation through Inhomogeneous Media," *Proc. Nat. Acad. Sci. USA*, Vol. 42, 1956, pp. 629–632.

§13.7. See

R. Bellman and T. Brown, "A Note on Invariant Imbedding and Generalized Semigroups," *J. Math. Anal. Appl.*, Vol. 9, 1964.

§13.8. See

R. Bellman, R. Kalaba, and G. M. Wing, "Invariant Imbedding and Mathematical Physics—I: Particle Processes," *J. Math. Phys.*, Vol. 1, 1960, pp. 280–308,

for an extensive discussion of models of this nature and many further references.

For detailed discussions of these questions, see

R. Bellman, "Scattering Processes and Invariant Imbedding," *J. Math. Anal. Appl.*, Vol. 23, 1968, pp. 254–268.

R. Bellman, "Invariant Imbedding and a Nonlinear Scattering Process," *J. Math. Anal. Appl.*

§13.14. For a detailed account of computational aspects, see the first books by Bellman–Kalaba *et al.*, referred to in Sec. 13.1.

§13.20. See

I. I. Kolodner, "Phase Shift of Solutions of Second Order Linear Ordinary Differential Equations and Related Problems," *J. Math. Anal. Appl.*, Vol. 4, 1962, pp. 422–439.

§13.23. See

R. Bellman and R. Kalaba, "Functional Equations, Wave Propagation and Invariant Imbedding," *J. Math. Mech.*, Vol. 8, 1959, pp. 683–702.

Chapter 14 THE THEORY OF ITERATION

14.1. Introduction

In this chapter we wish to provide an introduction to some of the basic ideas of the theory of the iteration of analytic functions. The subject is a fascinating one with ramifications in wide-ranging fields such as physics, engineering, and biology. Consequently, although we shall provide numerous references, we shall do no more than touch upon some highlights in the text.

Let us indicate a number of different analytic motivations here. In the first place, our results enable us to advance some stages further in the important study of nonlinear difference equations of the form

$$x_{n+1} = g(x_n) = Ax_n + \cdots, \qquad x_0 = c, \tag{14.1.1}$$

as well as in the closely related study of nonlinear differential equations of the form

$$\frac{dx}{dt} = h(x) = Ax + \cdots, \qquad x(0) = c, \tag{14.1.2}$$

where g and h are analytic in their arguments in a neighborhood of the origin and lack constant terms.

The groundwork for these investigations was laid in Chapter 4, Volume I, where some fundamental results of Poincaré, Lyapunov, and Perron were presented. In Chapter 12, these results concerning asymptotic behavior of solutions of (14.1.1) and (14.1.2) were supplemented in one essential way using invariant imbedding. In this chapter we provide more definitive results using certain basic canonical representations and relative invariants.

Secondly, we find that these representations play an important role in the study of equations of the foregoing type when the initial values are random variables. The solutions of the Abel–Schröder functional equation enable us in favorable cases to bypass difficult truncation problems posed by the conventional approach. We shall explore those matters of truncation in more detail in Chapter 15. We can regard the investigations in this chapter as a continuation of our persistent efforts in the area of closure of operations.

An integral part of a general theory of closure of operations is the determination of suitable canonical transformations. Thus, we may regard our present investigations as a natural continuation of the study of canonical transformations for matrices described in Chapter 2, Volume I. As might be expected, nonlinearity introduces many novel, complex, and interesting features. Consequently, we will touch only lightly on the multidimensional aspects.

Finally, we shall indicate some classes of partial differential equations that are connected with iteration in the study of different classes of processes, both those described by initial value processes and boundary-value problems. Here we again make contact with invariant imbedding and the theory of semigroups.

14.2. Iteration

In Chapter 12 we indicated the connection between the iteration of transformations and dynamical processes, an idea extensively developed by Poincaré, Hadamard, Birkhoff, Levi-Civita, and many others. Here we wish to restrict ourselves to the analytic case where the transformation is analytic in its arguments. To simplify initially, let us consider the one-dimensional case.

Consider then the transformation

$$h(u) = a_1 u + a_2 u^2 + \cdots, \tag{14.2.1}$$

where the power series converges in some interval $|u| < c_1$. If we further assume that $0 < |a_1| < 1$, it follows that there is an interval, $|u| \leqslant c_2 < c_1$ with the property that $h(u)$ is a shrinking transformation in this interval, i.e., $|h(u)| \leqslant c_2$ for $|u| \leqslant c_2$. In this interval, we are interested in studying the iterates

$$h(u), \quad h(h(u)), \dots . \tag{14.2.2}$$

This is equivalent to studying the behavior of the solution of the difference equation

$$u_{n+1} = h(u_n), \qquad u_0 = u. \tag{14.2.3}$$

We shall write

$$h^{(n)}(u) = h(h \cdots n \text{ times } (u)), \tag{14.2.4}$$

and call it the nth iterate of h. Clearly, a semigroup property holds

$$h^{(m+n)} = h^{(m)}(h^{(n)}) = h^{(n)}(h^{(m)}), \qquad h^{(0)} = u. \tag{14.2.5}$$

An important point is that each function $h^{(n)}$ is a power series in u,

$$h^{(n)}(u) = a_1^{(n)}u + a_2^{(n)}u^2 + \cdots + a_m^{(n)}u^m + \cdots. \tag{14.2.6}$$

The question we wish to study is that of the analytic behavior of $a_m^{(n)}$, and of $h^{(n)}(u)$ in general, as functions of n.

14.3. Abel–Schröder Functional Equation

Suppose that we could find a function $\phi(u)$ such that

$$\phi(h(u)) = b\phi(u) \tag{14.3.1}$$

where b is some constant. Then we would have

$$\phi(h^{(2)}) = b\phi(h) = b^2\phi(u), \tag{14.3.2}$$

and thus, inductively,

$$\phi(h^{(n)}) = b^n\phi(u). \tag{14.3.3}$$

Hence, there results the elegant formula

$$h^{(n)} = \phi^{-1}(b^n\phi(u)). \tag{14.3.4}$$

Observe then that the determination of $h^{(n)}$, a complex matter, has been replaced by the process of effecting the transformation ϕ, the multiplication b^n, and then the transformation ϕ^{-1}. This is usually a tremendous reduction in effort. Furthermore, (14.3.4) yields the desired information concerning the analytic structure of $h^{(n)}$.

The equation in (14.3.1), quite important in analysis, is called the Abel–Schröder functional equation and has been intensively studied.

14.4. Formal Analysis

Prior to a rigorous analysis, let us see whether there is a possibility of an analytic solution to (14.3.1) of the form

$$\phi(u) = u + b_2u^2 + \cdots. \tag{14.4.1}$$

It is clearly permissible to normalize by taking the coefficient of u to be 1, if it is nonzero, since any constant multiple of ϕ is also a solution. Substituting in (14.3.1), we see that

$$h(u) + b_2 h(u)^2 + \cdots + b_n h(u)^n + \cdots = b[u + b_2 u^2 + \cdots + b_n u^n + \cdots] \quad (14.4.2)$$

which upon equating coefficients yields

$$b = a_1 \quad (14.4.3)$$

and recurrence relations for $b_2, b_3, \ldots, b_n, \ldots$, which uniquely determine these coefficients in terms of $a_1, a_2, \ldots$. Hence, if an analytic $\phi(u)$ exists with $\phi'(0) = 1$, it is unique.

Exercises

1. Determine a general recurrence relation for b_n.
2. What about analytic solutions of (14.3.1) with $\Psi'(0) = 0$, $\Psi''(0) = 1$? What relation do they have to $\phi(u)$ above?

14.5. Koenig's Representation

The simplest existence proof for ϕ is one that we have essentially already given in the chapters on stability theory and invariant imbedding.

The difference equation

$$u_{n+1} = h(u_n) = a_1 u_n + \cdots, \qquad u_0 = c, \quad (14.5.1)$$

is one of Poincaré–Lyapunov–Perron type. If $| a_1 | < 1$ and $| c | \ll 1$, we know that

$$\lim_{n\to\infty} \frac{u_n}{a_1^n} = \phi(c) \quad (14.5.2)$$

exists and that $\phi(c)$ is analytic for $| c | \ll 1$.

An invariant imbedding argument now shows that

$$\phi(c) = \lim_{n\to\infty} \frac{u_n}{a_1^n} = \lim_{n\to\infty} \frac{u_{n+1}}{a_1^{n+1}} = \frac{1}{a_1} \lim \frac{u_{n+1}}{a_1^n} = \frac{\phi(h(c))}{a_1}, \quad (14.5.3)$$

since the initial condition for u_{n+1} at $n = 0$ is $u_1 = \phi(h(c))$.

This elegant representation of the function $\phi(c)$ is due to Koenigs.

Exercise

1. Write $h(u) = a_1 u(1 + g(u))$ for $|u| \ll 1$. Show that

$$\phi(c) = c \prod_{k=0}^{\infty} [1 + g(h^{(k)}(c))].$$

14.6. Majorization

A second proof of the existence of an analytic function satisfying the Abel–Schröder functional equation can be given using the classical technique of majorization due to Cauchy. Let $u(c)$ be analytic for $|c| < c_1$,

$$u(c) = \sum_{n=0}^{\infty} u_n c^n. \tag{14.6.1}$$

We write

$$u(c) \ll v(c), \tag{14.6.2}$$

and say that $v(c)$ *majorizes* $u(c)$ if $v(c)$ is another function analytic in the same region,

$$v(c) = \sum_{n=0}^{\infty} v_n c^n, \tag{14.6.3}$$

with the property that

$$|u_n| \leqslant v_n, \qquad n = 0, 1, 2, \ldots . \tag{14.6.4}$$

What is essential is that majorization is preserved under addition and multiplication, i.e.,

$$u_1 \ll v_1, \qquad u_2 \ll v_2 \tag{14.6.5}$$

imply that

$$\begin{aligned} &\text{(a)} \quad b_1 u_1 + b_2 u_2 \ll b_1 v_1 + b_1 v_2, \qquad \text{where} \quad b_1, b_2 \geqslant 0, \\ &\text{(b)} \quad u_1 u_2 \ll v_1 v_2 . \end{aligned} \tag{14.6.6}$$

Furthermore, it is a transitive operation,

$$u_1 \ll u_2, \quad u_2 \ll u_3 \qquad \text{implies} \quad u_1 \ll u_3 . \tag{14.6.7}$$

Finally, it is preserved by iteration,

$$u_1 \ll v_1, \quad u_2 \ll v_2 \qquad \text{imply} \quad u_1(u_2) \ll v_1(v_2).$$

We leave the proofs as exercises. Observe that the concept of majorization carries over to formal power series.

Returning to the functional equation

$$\phi(h) = a_1 \phi, \tag{14.6.8}$$

and noting the discussion in Sec. 14.4, it is easy to show inductively that if $0 < | a_1 | < 1$ and $h \ll g$, then $\phi \ll \Psi$, where Ψ is the function determined, formally at least, by the companion equation

$$\Psi(g) = | a_1 | \Psi. \tag{14.6.9}$$

The next step hinges upon the important observation that we can choose a majorizing function g for which (14.6.9) is readily solved. We begin with the fact that if $h(c)$ is analytic in $| c | < c_1$, we can find a function of the form

$$g(c) = \frac{| a_1 | c}{1 - kc}, \qquad k > 0, \tag{14.6.10}$$

which majorizes $h(c)$. Hence, let us consider the solution to the functional equation

$$\Psi\left(\frac{| a_1 | c}{1 - kc}\right) = | a_1 | \Psi(c). \tag{14.6.11}$$

We look for a solution of the simple form

$$\Psi(c) = \frac{c}{1 - bc}. \tag{14.6.12}$$

Substituting, we see that

$$\frac{| a_1 | c/(1 - kc)}{1 - b | a_1 | c/(1 - kc)} = \frac{| a_1 | c}{1 - (k + b | a_1 |)c} = \frac{| a_1 | c}{1 - bc}. \tag{14.6.13}$$

Hence, we can take

$$b = k/(1 - | a_1 |), \tag{14.6.14}$$

so that $k + b | a_1 | = b$ and thus obtain the desired majorant solution. Since it converges for $| c | \ll 1$, it follows that the formal solution obtained via the procedure in Sec. 4 converges in the same interval.

Exercise

1. Using the same type of argument, consider the differential equation $u' = g(u)$, $u(0) = c$, where g is analytic for $|u| < c$. Show that a majorant solution is given by a solution of the quadratic equation $u - (bu^2/2) = at + c$. Which solution? What is the radius of convergence?

14.7. Refined Asymptotic Behavior

Once the function $\phi(c)$ has been obtained, we can immediately use it to obtain a much more accurate portrayal of the asymptotic behavior of the sequence $\{u_n\}$ determined by

$$u_{n+1} = h(u_n), \qquad u_0 = c. \tag{14.7.1}$$

In place of the previous relation

$$u_n \sim a_1{}^n \phi(c), \tag{14.7.2}$$

we can now write

$$\phi(u_{n+1}) = \phi(h(u_n)) = a_1\phi(u_n). \tag{14.7.3}$$

Iterating this relation, we obtain

$$\phi(u_n) = a_1{}^n\phi(c), \qquad u_n = \phi^{-1}(a_1{}^n\phi(c)) = a_1{}^n\phi(c) + d_2a_1^{2n}\phi(c)^2 + \cdots. \tag{14.7.4}$$

Exercises

1. Obtain corresponding results by converting $u_{n+1} = h(u_n)$, $u_0 = c$ into a nonlinear sum equation. What are the advantages and disadvantages of each approach?
2. Use (14.7.4) to obtain the power series in c for u_n. Compare with the procedure of Exercise 1.

14.8. Stochastic Effects

So far we have considered a purely deterministic process. Let us now consider some ways in which we can introduce stochastic features. There are three immediate options. We can first of all permit the transformation at each stage to be stochastic. Secondly, we can take the initial value to be stochastic. Thirdly, we can apply the transformation at stochastic times. Fourthly, of course, we can combine various possibilities. We shall discuss the first three in turn.

Let us note also that we can introduce dynamic programming concepts and treat stochastic control processes. We shall not, however, pursue this route here.

14.9. Stochastic Iteration

A first important extension of the classical iteration process is obtained by supposing that a stochastic transformation is applied at each stage. As a quite simple but important example of this, consider a nonlinear transformation of the type

$$h(u) = r_1 u + r_2 u^2 + \cdots + r_k u^k + \cdots, \tag{14.9.1}$$

where the coefficients $r_1, r_2, \ldots$ are now random variables. For the sake of ready exposition, consider the simple case where with probability p we employ the transformation

$$T_1(u) = a_1 u + a_2 u^2 + \cdots, \tag{14.9.2}$$

and with probability $q = 1 - p$ the transformation

$$T_2(u) = b_1 u + b_2 u^2 + \cdots, \tag{14.9.3}$$

where the a_i and b_i, $i = 1, 2, \ldots$, are fixed parameters.
Each of these power series is assumed convergent for $|u|$ sufficiently small, and we suppose that $0 < a_1, b_1 < 1$, whence $0 < pa_1 + qb_1 < 1$.

If we are interested only in terminal behavior, say the expected value of $g(u_N)$, then setting $f_N(c) = E[g(u_N)]$, where E denotes the expected value, we readily obtain the recurrence relation

$$f_N(c) = pf_{N-1}(T_1(c)) + qf_{N-1}(T_2(c)), \qquad f_0(c) = c. \tag{14.9.4}$$

It is of interest to determine the asymptotic behavior of $f_N(c)$ as $N \to \infty$.

In the classical case of iteration, $p = 1$, $q = 0$, where (14.9.4) reduces to $f_N(c) = f_{N-1}(T_1(c))$, the result of Schröder, discussed above enables the asymptotic behavior of $f_N(c)$ to be readily ascertained.

In the case of the equation in (14.9.4), analogous results hold, but neither as simple, nor as complete. That for $|c|$ small,

$$\lim_{N\to\infty} \frac{f_N(c)}{(pa_1 + qb_1)^N} = \phi(c) \tag{14.9.5}$$

where $\phi(c)$ is the analytic solution of the equation

$$(pa_1 + qb_1)\,\phi(c) = p\phi\,(T_1(c)) + q\phi\,(T_2(c)), \qquad \phi'(0) = 1, \quad (14.9.6)$$

may be established without difficulty. We leave this as an exercise.

This result, however, merely determines the leading term in the asymptotic expansion of $f_N(c)$. To obtain the next term, we set

$$f_N(c) = (pa_1 + qb_1)^N \phi(c) + g_N(c). \quad (14.9.7)$$

By virtue of (14.9.6), $g_N(c)$ satisfies the same equation as $f_N(c)$, with the difference that

$$g_N'(0) = 0. \quad (14.9.8)$$

Hence, we look for a solution of the equation in (14.9.4) with a different normalization, namely $g_1(c) = c^2 + \cdots$, whence $g_N(c) = u_N{}^2 + \cdots$, and a solution of the new functional equation,

$$(pa_1{}^2 + qb_1{}^2)\phi_2(c) = p\phi_2[T_1(c)] + q\phi_2[T_2(c)], \qquad \phi_2''(0) = 2. \quad (14.9.9)$$

Then, it is easy to show that

$$g_N(c) \sim (pa_1{}^2 + qb_1{}^2)^N \phi_2(c), \quad (14.9.10)$$

as $N \to \infty$. Combining (14.9.7) and (14.9.10) we have

$$f_N(c) = (pa_1 + qb_1)^N \phi(c) + (pa_1{}^2 + qb_1{}^2)^N \phi_2(c) + O[(pa_1{}^2 + qb_1{}^2)^N]. \quad (14.9.11)$$

We may then continue, step-by-step, to find the higher-order terms. It is clear from our assumptions that

$$pa_1{}^2 + qb_1{}^2 < pa_1 + qb_1\,. \quad (14.9.12)$$

14.10. Stochastic Matrices

Another case, a very important one, is that where the transformation is linear. This means that we are dealing with the product of random matrices, or equivalently with the linear difference equation

$$x_{n+1} = R_n x_n\,, \qquad x_0 = c, \quad (14.10.1)$$

where the components of R_n are independent random variables. Let us consider the two-dimensional case and write this equation in terms of components and elements,

$$u_{n+1} = z_{11}^{(n)} u_n + z_{12}^{(n)} v_n,$$
$$v_{n+1} = z_{21}^{(n)} u_n + z_{22}^{(n)} v_n. \tag{14.10.2}$$

The $z_{ij}^{(n)}$ are independent random variables and our desire is to determine the moments $E(u_{n+1}^k)$, $E(v_{n+1}^k)$. We shall find that Kronecker matrices play a fundamental role.

The case $k = 1$ is immediate. We have

$$E(u_{n+1}) = E(z_{11}^{(n)}) E(u_n) + E(z_{12}^{(n)}) E(v_n),$$
$$E(v_{n+1}) = E(z_{21}^{(n)}) E(u_n) + E(z_{22}^{(n)}) E(v_n), \tag{14.10.3}$$

whence

$$E(x_{n+1}) = E(R)^{n+1}c. \tag{14.10.4}$$

The case $k = 2$ is a bit more difficult and illustrates the general idea. We have from (14.10.2)

$$u_{n+1}^2 = (z_{11}^{(n)})^2 u_n^2 + 2z_{11}^{(n)} z_{12}^{(n)} u_n v_n + (z_{12}^{(n)})^2 v_n^2,$$
$$v_{n+1}^2 = (z_{21}^{(n)})^2 u_n^2 + 2z_{21}^{(n)} z_{22}^{(n)} u_n v_n + (z_{22}^{(n)})^2 v_n^2, \tag{14.10.5}$$
$$u_{n+1}v_{n+1} = z_{11}^{(n)} z_{21}^{(n)} u_n^2 + (z_{11}^{(n)} z_{22}^{(n)} + z_{12}^{(n)} z_{21}^{(n)}) u_n v_n + z_{12}^{(n)} z_{22}^{(n)} v_n^2.$$

We see then that

$$\begin{pmatrix} E(u_{n+1}^2) \\ E(u_{n+1}v_{n+1}) \\ E(v_{n+1}^2) \end{pmatrix} = E(R^{[2]}) \begin{pmatrix} E(u_n^2) \\ E(u_n v_n) \\ E(v_n^2) \end{pmatrix} \tag{14.10.6}$$

where $R^{[2]}$ is the Kronecker square of R.

Exercises

1. Show how to obtain the kth moments in terms of the Kronecker kth power of R.

2. Consider the case where

$$R_m = \begin{pmatrix} 1 & \Delta \\ -\Delta r_n & 1 \end{pmatrix}$$

where the r_n are independent random variables. Obtain the asymptotic form of the first and second moments as $N \to \infty$ and the asymptotic behavior of the $E(u_n)$, $E(u_n^2)$ as $\Delta \to 0$.

3. Consider the case where r_n is a Markov process. See

R. Bellman, T. T. Soong, and R. Vasudevan, "On Moment Behavior of Stochastic Difference Equations," *J. Math. Anal. Appl.* (to appear).

14.11. Random Initial Values

Let us next consider the case where c, the initial value, is a random variable. Using the relation

$$u_{n+1} = a_1 u_n + a_2 u_n^2 + \cdots + a_k u_n^k + \cdots, \tag{14.11.1}$$

we readily obtain

$$E(u_{n+1}) = a_1 E(u_n) + a_2 E(u_n^2) + \cdots + a_k E(u_n^k) + \cdots, \tag{14.11.2}$$

where E denotes the expected value taken with respect to the random variable c. Raising both sides of (14.11.1) to the second, third, etc. power, we can obtain corresponding results for the higher moments,

$$\begin{aligned} E(u_{n+1}^2) &= a_1^2\, E(u_n^2) + 2a_1 a_2\, E(u_n^3) + \cdots, \\ E(u_{n+1}^3) &= a_1^3\, E(u_n^3) + \cdots. \end{aligned} \tag{14.11.3}$$

We obtain in this way an infinite sequence of linear recurrence relations expressing $E(u_{n+1}^k)$, $k = 1, 2, \ldots$, in terms of $E(u_n^r)$, $r \geqslant k$. If we wish in this fashion to obtain an expression for $E(u_n)$, $E(u_n^2), \ldots$, in terms of $E(c)$, $E(c^2), \ldots$, we must employ a closure technique, which is to say, some type of truncation. We shall discuss systematic procedures for this in Chapter 15.

Using the canonical representation of Sec. 14.7, we can, however, proceed in a far simpler fashion. Since

$$\phi(u_n) = a_1^n\, \phi(c), \tag{14.11.4}$$

we have

$$E(\phi(u_n)^k) = (a_1^k)^n\, E(\phi(c)^k), \qquad k = 1, 2, \ldots. \tag{14.11.5}$$

Hence, if we write

$$u_n = \phi^{-1}(\phi(u_n)) = \phi(u_n) + b_2\,\phi(u_n)^2 + \cdots, \tag{14.11.6}$$

we have

$$E(u_n) = a_1{}^n\, E(\varphi(c)) + b_2 a_1^{2n}\, E(\phi(c)^2) + \cdots, \tag{14.11.7}$$

all under the assumption that the power series expansions remain valid. This is certainly the case, provided that c only assumes values in a region $|\, c \,| \leqslant c_0 \ll 1$.

14.12. Imbedding

Before considering the third case of random times, let us discuss some matters of a different nature.

Consider the differential equation

$$\frac{du}{dt} = g(u) = b_1 u + b_2 u^2 + \cdots, \qquad u(0) = c, \tag{14.12.1}$$

where $g(u)$ is analytic for $|\, u \,| < u_0$. Then, if $|\, c \,| \ll 1$, we can continue the solution to $t = 1$ and assert that

$$u(1) = h(c) \tag{14.12.2}$$

is an analytic function of c in $|\, c \,| \ll 1$,

$$h(c) = a_1 c + a_2 c^2 + \cdots. \tag{14.12.3}$$

If $b_1 < 0$, then $a_1 < 1$. A question of some interest is whether every analytic transformation of the form appearing in (14.12.3) can be considered to have arisen in this fashion.

Another way of putting this is the following: Given the transformation of (14.12.3), with $|\, a_1 \,| < 1$, can we imbed it in a family of transformations $u(c, t)$, $t \geqslant 0$, having the property that

$$\begin{aligned} &\text{(a)}\quad u(c, 0) = c, \\ &\text{(b)}\quad u(c, 1) = h(c), \\ &\text{(c)}\quad u(c, t + s) = u(u(c, t), s) \qquad t, s \geqslant 0\,? \end{aligned} \tag{14.12.4}$$

We see that the family is a semigroup in t.

With the aid of the canonical representation

$$h^{(n)} = \phi^{-1}(a_1{}^n\phi) \tag{14.12.5}$$

for the nth iterate, the answer is immediate. Introduce the function

$$u(c, t) = \phi^{-1}\,[a_1{}^t\phi(c)]. \tag{14.12.6}$$

It is clear all three conditions of (14.12.4) are readily satisfied. Furthmore

$$\frac{d}{dt}\,\phi(u(c, t)) = \frac{d}{dt}(a_1{}^t\phi(c)), \qquad \phi'(u(c, t))u_t = a_1{}^t\phi(c)\log a_1, \tag{14.12.7}$$

whence

$$\frac{du}{dt} = \frac{\phi(u)\log a_1}{\phi'(u)}, \qquad u(0) = c, \tag{14.12.8}$$

is the desired differential equation.

Exercises

1. If $a_1, a_2, \ldots, \geqslant 0$, when are the coefficients in $u(c, t)$, as a power series in c, nonnegative?

14.13. Differential Equations

A continuous version of iteration is thus obtained by considering the function of two variables $u(c, t)$ satisfying the fundamental semigroup condition of (14.12.4c),

$$u(c, s + t) = u(u(c, s), t), \qquad s, t \geqslant 0, \tag{14.13.1}$$

with $u(c, 0) = c$. We obtain this from a continuous version of a difference equation, namely a differential equation of the form

$$\frac{du}{dt} = g(u), \qquad u(0) = c. \tag{14.13.2}$$

If we suppose that the solution of (14.13.2) is unique for $t \geqslant 0$, we can derive (14.13.1) by considering the solution at $s + t$ as the result of a solution that is observed at time s, in state $u(c, s)$, and continued for another interval of length t. Conversely, given (14.13.1) and the differentiability with respect to s and t, we obtain (14.13.2) as a limiting form as s approaches zero.

Since (14.13.2) can be integrated explicitly, there is no difficulty in finding a relative invariant, i.e., a function $\psi(u)$ such that

$$\psi(u)' = b\psi(u), \tag{14.13.3}$$

when one exists. Let us, however, consider a more general method with the multidimensional case in mind.

If $g(u)$ has the form

$$g(u) = -b_1 u + h(u) \tag{14.13.4}$$

where $h(u) = O(u^2)$ as $u \to 0$, $b_1 > 0$, we know from Poincaré–Lyapunov theory (see Chapter 3, Volume I) that u, as given by (14.13.2), exists for all $t \geqslant 0$, and

$$\lim_{t\to\infty} ue^{b_1 t} = \varphi(c), \tag{14.13.5}$$

where $\varphi(c)$ is analytic for $|c| \ll 1$, provided that $h(u)$ is analytic for $|u| \ll 1$. For the case where g is analytic in u, we have already given (in Sec. 13.18) a prescription a la invariant imbedding for the determination of $\varphi(c)$ in the scalar, and more importantly, in the multidimensional case. The determination of relative invariants in the multidimensional case is far more complex, as we shall see in a moment.

14.14. Associated Partial Differential Equation

Let $u(c, t)$ be the solution of (14.13.2). Then,

$$u(c, t) = u(c + \Delta g(c), t - \Delta) \tag{14.14.1}$$

to terms in Δ^2 for $\Delta \ll 1$. Hence u satisfies the partial differential equation

$$0 = g(c)\frac{\partial u}{\partial c} - \frac{\partial u}{\partial t}, \tag{14.14.2}$$

with $u(c, 0) = c$. In the multidimensional case, where the associated differential equation is

$$\frac{dx}{dt} = g(x), \qquad x(0) = c, \tag{14.14.3}$$

each of the components of x satisfies the equation

$$0 = (\operatorname{grad} f, g(c)) - \frac{\partial f}{\partial t}. \tag{14.14.4}$$

More generally, any differentiable scalar function of x satisfies this equation. The gradient is taken with respect to c.

The characteristic functions of

$$(\text{grad}\, f, g(c)) = \lambda g(c) \tag{14.14.5}$$

are relative invariants of the original differential equation.

14.15. Multidimensional Case

Let us now consider the more interesting and difficult case of a vector transformation

$$x_{n+1} = Ax_n + g(x_n), \qquad x_0 = c, \tag{14.15.1}$$

where we suppose that the characteristic roots of A, $(\lambda_1, \lambda_2, \dots, \lambda_N)$, are distinct and less than one in absolute value, and, as usual, that all of the components of $g(x)$ are power series in the components of x lacking constant and first-order terms. Write

$$x_n = Ty_n, \tag{14.15.2}$$

where T has been chosen so that A is reduced to diagonal form

$$T^{-1}AT = \begin{bmatrix} \lambda_1 & & & 0 \\ & \lambda_2 & & \\ & & \ddots & \\ 0 & & & \lambda_N \end{bmatrix} = \Lambda. \tag{14.15.3}$$

Then, substituting in (14.15.1), we obtain the simpler relation

$$y_{n+1} = \Lambda y_n + T^{-1}g(Ty_n) = \Lambda y_n + h(y_n), \qquad y_0 = T^{-1}c = c'. \tag{14.15.4}$$

If $g(x)$ is such that all components are power series in the components of x lacking constant and first-degree terms, then $h(y)$ has the same property as a function of y.

To simplify the notation, let us consider the two-dimensional case where the complexity of the situation is already apparent, as will be indicated below. Write

$$\begin{aligned} f(x_1, x_2) &= \rho x_1 + \sum_{k+l\geqslant 2} a_{kl}x_1^k x_2^l, \\ g(x_1, x_2) &= \sigma x_2 + \sum_{k+l\geqslant 2} b_{kl}x_1^k x_2^l, \end{aligned} \tag{14.15.5}$$

valid for $|x_1|$, $|x_2| \ll 1$, where $0 < \rho$, $\sigma < 1$. We take $\rho > \sigma$ to be specific and write

$$u_{n+1} = f(u_n, v_n), \quad u_0 = c_1, \qquad v_{n+1} = g(u_n, v_n), \quad v_0 = c_2, \tag{14.15.6}$$

where $|c_1|$ and $|c_2|$ are small. The Perron theorem, discussed in Chapter 4, Volume I, tells us that we have

$$\lim_{n\to\infty} \frac{u_n}{\rho^n} = \varphi(c_1, c_2), \qquad \lim_{n\to\infty} \frac{v_n}{\rho^n} = \psi(c_1, c_2), \tag{14.15.7}$$

where φ and ψ are analytic functions of c_1 and c_2. It is clear, as before, that we have the functional equations

$$\varphi(f, g) = \rho\varphi(c_1, c_2), \qquad \psi(f, g) = \rho\psi(c_1, c_2). \tag{14.15.8}$$

Consequently, we appear to have our two relative invariants. Unfortunately, it is easy to see that $\psi(c_1, c_2) \equiv 0$, which means that this procedure furnishes only one function, ϕ.

To overcome this difficulty, it is tempting to parallel the preceding analysis and to introduce the function

$$\lim_{n\to\infty} \frac{v_n}{\sigma^n} = \psi(c_1, c_2). \tag{14.15.9}$$

If $\psi(c_1, c_2)$ exists, it yields a function satisfying the functional equation

$$\psi(f, g) = \sigma\psi(c_1, c_2). \tag{14.15.10}$$

As we shall see, this is basically a sound idea. There are, however, some serious difficulties to overcome and the procedure has to be modified substantially in certain cases.

14.16. A Counterexample

In order to show the kind of difficulty that can arise, let us consider the particular case where

$$f(x_1, x_2) = \rho x_1, \qquad g(x_1, x_2) = \sigma x_2 + x_1^2. \tag{14.16.1}$$

It is easily shown by induction that

$$u_n = \rho^n c_1, \qquad v_n = \sigma^n c_2 + [\sigma^{n-1} + \sigma^{n-2}\rho^2 + \cdots + \rho^{2(n-1)}]c_1^2, \tag{14.16.2}$$

where we are employing the notation of the previous section. Thus, we have explicitly

$$\phi(c_1, c_2) = \lim_{n\to\infty} \frac{u_n}{\rho^n} = c_1$$
$$\psi(c_1, c_2) = \lim_{n\to\infty} \frac{v_n}{\sigma^n} = c_2 + \frac{c_1^2}{\sigma} \lim_{n\to\infty} \left(1 + \frac{\rho^2}{\sigma} + \cdots + \left(\frac{\rho^2}{\sigma}\right)^{n-1}\right). \tag{14.16.3}$$

This last limit exists only for $|\rho^2/\sigma| < 1$. Hence, the general procedure of (14.15.9) fails. Yet, if undeterred by this, we use the method of undetermined coefficients to determine two analytic functions ϕ and ψ satisfying the functional equations

$$\phi(\rho c_1, \sigma c_2 + c_1^2) = \rho\phi(c_1, c_2), \qquad \psi(\rho c_1, \sigma c_2 + c_1^2) = \sigma\psi(c_1, c_2), \tag{14.16.4}$$

both vanishing at the origin, and normalized to have the forms $\phi(c_1, c_2) = c_1 + \cdots$, $\psi(c_1, c_2) = c_2 + \cdots$. We readily find that

$$\phi(c_1, c_2) = c_1, \qquad \psi(c_1, c_2) = c_2 + \frac{c_1^2}{\sigma - \rho^2}, \tag{14.16.5}$$

provided that $\sigma \neq \rho^2$.

Comparing (14.16.3) and (14.16.5), we see that an easy explanation of the foregoing lies in the fact that the expression $\sigma/\sigma - \rho^2)$ can be considered to be the "sum" of the series

$$1 + \frac{\rho^2}{\sigma} + \cdots + \left(\frac{\rho^2}{\sigma}\right)^n + \cdots, \tag{14.16.6}$$

by means of any of a number of summability methods, provided that $\sigma \neq \rho^2$.

Consideration of the functions

$$f(x_1, x_2) = \rho x_1, \qquad g(x_1, x_2) = \sigma x_2 + \sum_{k=2}^{\infty} c_k x_1^k, \tag{14.16.7}$$

with the associated relative invariants

$$\phi(c_1, c_2) = c_1, \qquad \psi(c_1, c_2) = c_2 - \sum_{k+2}^{\infty} \frac{c_k c_1^k}{\rho^k - \sigma}, \tag{14.16.8}$$

shows that the condition $\sigma \neq \rho^k$, $k = 2, 3, \ldots$, is necessary for the existence of analytic solutions of the basic functional equations having the prescribed form. As we shall indicate below, it is also sufficient, in the case where $1 > |\rho| > |\sigma > 0$.

14.17. Statement of Result

It can be shown in a number of different ways that $1 > |\rho| > |\sigma| > 0$, $\sigma \neq \rho^n$, is a sufficient condition for the existence of two analytic functions of the form

$$\phi(c_1, c_2) = c_1 + \cdots, \qquad \psi(c_1, c_2) = c_2 + \cdots, \tag{14.17.1}$$

such that

$$\phi(f, g) = \rho\varphi(c_1, c_2), \qquad \psi(f, g) = \sigma\psi(c_1, c_2), \tag{14.17.2}$$

for $|c_1|$, $|c_2|$ sufficiently small. Thus solving (14.17.2) for f and g under the assumption that $|c_1|$, $|c_2|$ are small, we can write

$$f(c_1, c_2) = F(\rho\varphi, \sigma\psi), \qquad g(c_1, c_2) = G(\rho\phi, \sigma\psi), \tag{14.17.3}$$

where F and G are again analytic for $|c_1|, |c_2| \ll 1$. Hence, if

$$u_{n+1} = f(u_n, v_n), \quad u_0 = c_1, \qquad v_{n+1} = g(u_n, v_n), \quad v_0 = c_2, \tag{14.17.4}$$

we see that

$$u_n = F(\rho^n\phi, \sigma^n\psi), \qquad v_n = G(\rho^n\varphi, \sigma^n\psi), \tag{14.17.5}$$

an elegant representation which completely resolves the question of asymptotic behavior. References to a number of different derivations of this result will be found at the end of the chapter; see also the following section.

We can use this result as before to study the behavior of u_n and v_n when c_1 and c_2 are random variables.

14.18. Differential Equations

Analogous results hold for nonlinear differential equation of the form

$$\frac{dx}{dt} = Ax + g(x), \qquad x(0) = c, \tag{14.18.1}$$

where the characteristic roots of A are distinct and have negative real parts and $g(x)$ is as above. Let once again

$$x = Ty, \tag{14.18.2}$$

where

$$T^{-1}AT = \begin{bmatrix} \lambda_1 & & & 0 \\ & \lambda_2 & & \\ & & \ddots & \\ 0 & & & \lambda_N \end{bmatrix}. \tag{14.18.3}$$

Then

$$\frac{dy}{dt} = \Lambda y + T^{-1}g(Ty), \qquad y(0) = T^{-1}c, \tag{14.18.4}$$
$$= \Lambda y + h(y), \qquad y(0) = c',$$

where $h(y)$ lacks constant and first-order terms. It can be shown that if no relation of the form

$$\sum_{k=1}^{N} m_k \lambda_k = 0 \tag{14.18.5}$$

holds, where the m_k are rational integers, then we can find analytic functions $\phi_i(c_1, c_2, \ldots, c_N)$, $i = 1, 2, \ldots$, such that we have a complete set of relative invariants of the form

$$\phi_i(x_1, x_2, \ldots, x_N) = e^{\lambda_i t} \phi_i(c_1, c_2, \ldots, c_N), \qquad i = 1, 2, \ldots, N. \tag{14.18.6}$$

Without going into the details which are onerous, let us indicate briefly why we expect the result to be true. As we know from the Poincaré–Lyapunov theorem, if (14.18.4) holds, Λ is a stability matrix and $\| c \| \ll 1$ then the solution exists for $t \geqslant 0$. Writing

$$y = e^{\Lambda t}c + \int_0^t e^{\Lambda(t-t_1)} h(y(t_1))\, dt_1, \tag{14.18.7}$$

we see, following Lyapunov, that y may be expressed as a power series in $e^{\lambda_1 t}, e^{\lambda_2 t}, \ldots, e^{\lambda_N t}$, if the condition of (14.18.5) holds. As a result of the integrations to be carried out in the second term of the right-hand side of (14.18.7), there are no "resonances," i.e., no secular terms enter. Hence, if $y_1, y_2, \ldots, y_N$ denote the components of y, we have N relations of the form

$$y_i = g_i(e^{\lambda_1 t}, e^{\lambda_2 t}, \ldots, e^{\lambda_N t}), \qquad i = 1, 2, \ldots, N. \tag{14.18.8}$$

Solving these N equations for the quantities $e^{\lambda_1 t}, e^{\lambda_2 t}, \ldots, e^{\lambda_N t}$, we have the result

$$e^{\lambda_i t} = \phi_i(y_1, y_2, \ldots, y_N), \qquad i = 1, 2, \ldots, N. \tag{14.19.9}$$

The functions φ_i thus constitute a desired set of relative invariants, since clearly

$$\frac{d\varphi_i}{dt} = \lambda_i \varphi_i . \tag{14.18.10}$$

14.19. Commensurable Characteristic Roots

Certain "resonance" effects occur when the characteristic roots are commensurate. To see this, consider the two-dimensional system

$$\frac{du}{dt} = -u, \qquad u(0) = c_1,$$
$$\frac{dv}{dt} = -2v + u^2, \qquad v(0) = c_2. \tag{14.19.1}$$

We have

$$u = c_1 e^{-t}, \qquad v = c_2 e^{-2t} + c_1{}^2\, te^{-2t}, \tag{14.19.2}$$

as is easily verified. Note then that

$$\frac{d}{dt}(v - c_1{}^2\, te^{-2t}) = -2c_2 e^{-2t} \tag{14.19.3}$$

or if

$$\phi_2(u, v) = v - tu^2, \tag{14.19.4}$$

that

$$\frac{d\phi_2}{dt} = -2\phi_2, \qquad \phi_2(0) = c_2. \tag{14.19.5}$$

Hence,

$$\phi_1(u, v) = u, \qquad \phi_2(u, v) = v - tu^2 \tag{14.19.6}$$

constitute a set of relative "invariants."

The case of commensurable characteristic roots can thus be treated by allowing time-varying relative invariants. From the standpoint of invariance, this is certainly not very satisfying. However, with the object of providing a useful representation for the asymptotic behavior of the solution and for a treatment of the stochastic case, it is well suited to our purposes.

To avoid the t-dependence and thus obtain more respectable invariants, it is easy to see we can use the function

$$\phi_2(u, v) = v + u^2 \log u. \tag{14.19.7}$$

In the general case we can employ functions of the form

$$\phi_1(u, v) = u + \sum_{n+k \geqslant 2} a_{mnk} (\log u)^m u^n v^k, \tag{14.19.8}$$

$$\phi_2(u, v) = v + \sum_{n+k \geqslant 2} b_{mnk} (\log u)^m u^n v^k. \tag{14.19.9}$$

The subject is clearly one of some complexity and we shall therefore not pursue it any further here.

14.20. Two-point Boundary Values

Let us now briefly indicate that there is a corresponding set of results associated with two-point boundary value problems. Consider the two-dimensional system

$$u' = g(u, v), \quad u(a) = c_2, \qquad v' = h(u, v), \quad v(0) = c_1, \tag{14.20.1}$$

where $a > 0$. As we have previously discussed, we can intuitively think of this set of equations as describing the internal left- and right-hand fluxes respectively for a finite rod, $0 \leqslant t \leqslant a$ (Fig. 14.1), where c_1 and c_2 are external incident fluxes.

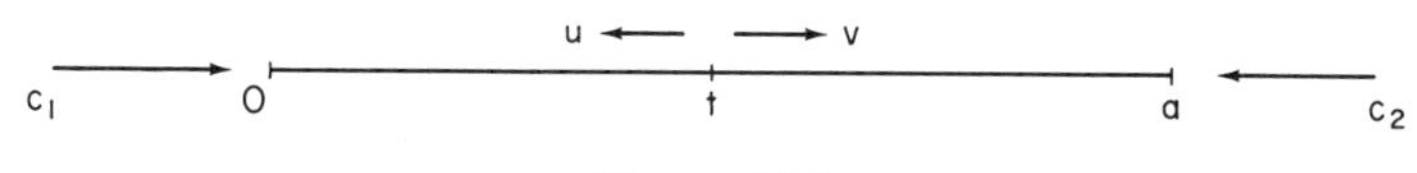

Figure 14.1

Let us use the approach of invariant imbedding discussed in the last chapter. We suppose first of all that there exists a unique solution of (14.20.1) for $a > 0$, $-\infty < c_1, c_2 < \infty$ (which is often not the case for arbitrarily large a and in any case not easy to establish), and introduce the functions

$$\begin{aligned} u(0) &= R(c_1, c_2, a), &&\text{the "reflected" flux,} \\ v(a) &= T(c_1, c_2, a), &&\text{the "transmitted" flux.} \end{aligned} \tag{14.20.1}$$

Frequently, only the physical region $0 \leqslant c_1, c_2 < \infty$ is of interest.

Examining Fig. 14.1, we have the relations

$$R(c_1, c_2, a) = R(c_1, u(t), t), \qquad v(t) = T(c_1, u(t), t), \tag{14.20.3}$$

concentrating upon the interval $[0, t]$. Similarly, the interval $[t, a]$ yields the relations

$$T(c_1, c_2, a) = T(v(t), c_2, a - t), \qquad u(t) = R(v(t), c_2, a - t). \tag{14.20.4}$$

Our aim is to show, as in the initial value case, that we can obtain the functions $R(c_1, c_2, N)$, and $T(c_1, c_2, N)$ for $a = N$, an integer, in terms of the functional values for $N = 1$.

To this end, introduce the basic functions,

$$r(c_1, c_2) = R(c_1, c_2, 1), \qquad t(c_1, c_2) = T(c_1, c_2, 1). \tag{14.20.5}$$

From the foregoing relations (14.20.4), with $a = 2$, $t = 1$, we have

$$\begin{aligned} &(\text{a}) \quad R(c_1, c_2, 2) = r(c_1, u(1)), \\ &(\text{b}) \quad v(1) = t(c_1, u(1)), \\ &(\text{c}) \quad T(c_1, c_2, 2) = t(v(1), c_2), \\ &(\text{d}) \quad u(1) = r(v(1), c_2). \end{aligned} \tag{14.20.6}$$

From (*b*) and (*d*), we obtain relations of the form

$$u(1) = w_1(c_1, c_2), \qquad v(1) = z_1(c_1, c_2). \tag{14.20.7}$$

Using (14.20.6a) and (14.20.6c), we obtain the desired relations,

$$\begin{aligned} R(c_1, c_2, 2) &= r(c_1, w_1(c_1, c_2)), \\ T(c_1, c_2, 2) &= t(z_1(c_1, c_2), c_2). \end{aligned} \tag{14.20.8}$$

Let us now consider the case where $a = N$. Using $t = 1$, we have from (14.20.6),

$$v(1) = t(c_1, u(1)), \qquad u(1) = R(v(1), c_2, N - 1). \tag{14.20.9}$$

Keep in mind that here $u(1) = u_N(1)$, $v(1) = v_N(1)$. These relations determine $u(1)$ and $v(1)$, namely

$$u(1) = w_{N-1}(c_1, c_2), \qquad v(1) = z_{N-1}(c_1, c_2). \tag{14.20.10}$$

Returning to (14.20.6), we have

$$\begin{aligned} R(c_1, c_2, N) &= r(c_1, w_{N-1}(c_1, c_2)), \\ T(c_1, c_2, N) &= t(z_{N-1}(c_1, c_2), c_2). \end{aligned} \tag{14.20.11}$$

Thus, the basic functions $r(c_1, c_2)$, $t(c_1, c_2)$ can be used to generate the corresponding functions for an arbitrary integral length.

14.21. Constant Right-hand Incident Flux

The results simplify greatly if we assume that the right-hand flux is constant, say $c_2 = 0$.

Let

$$f_N(c_1) = \text{the value of } u(0) \quad \text{for} \quad a = N. \tag{14.21.1}$$

Then we obtain the simpler functional equations

$$f_N(c_1) = r(c_1, u(1)), \qquad u(1) = f_{N-1}(v(1)), \quad v(1) = t(c_1, u(1)). \tag{14.21.2}$$

This provides the desired relations between f_N and f_{N-1}, upon eliminating $u(1)$ and $v(1)$ from the three equations. Again, $u(1) \equiv u_N(1)$, $v(1) \equiv v_N(1)$.

A most important case is that where g and h are analytic about $u = v = 0$, having the power series expansions

$$g(u, v) = a_1 u + a_2 v + \cdots, \qquad h(u, v) = b_1 u + b_2 v + \cdots. \tag{14.21.3}$$

In this case, we can set

$$\begin{aligned} f_N(c_1) &= r_{1N} c_1 + r_{2N} c_1^2 + \cdots, \\ u(1) &= u_{1N} c_1 + u_{2N} c_1^2 + \cdots, \\ v(1) &= v_{1N} c_1 + v_{2N} c_1^2 + \cdots \end{aligned} \tag{14.21.4}$$

and use (14.21.2) to obtain recurrence relations for the triple (r_{iN}, u_{iN}, v_{iN}) in terms of $(r_{i,N-1}, u_{i,N-1}, v_{i,N-1})$. The expansions will be valid for $|c_1|$ small.

The case where $N = \infty$ is particularly simple, since (14.21.2) reduces to

$$f(c_1) = r(c_1, u(1)), \qquad u(1) = f(v(1)), \quad v(1) = t(c_1, u(1)), \tag{14.21.5}$$

upon setting $f(c_1) = f_\infty(c_1)$.

Exercises

1. Determine the explicit forms of r_{1N}, u_{1N}, v_{1N}.
2. Determine the recurrence relations for r_{iN}, u_{iN}, v_{iN}.

14.22. Associated Partial Differential Equation

To derive an associated partial differential equation, we write

$$v(a - \Delta) = T(c_1, u(a - \Delta), a - \Delta), \tag{14.22.1}$$

where $\Delta \ll 1$. Then

$$\begin{aligned} v(a-\Delta) &= v(a) - \Delta v'(a) + \cdots, \\ &= T(c_1, c_2, a) - \Delta h(c_2, T) + \cdots, \\ u(a-\Delta) &= u(a) - \Delta u'(a) + \cdots, \qquad (14.22.2) \\ &= c_2 - \Delta g(c_2, T) + \cdots \\ T(c_1, u(a-\Delta), a-\Delta) &= T(c_1, c_2, a) - \Delta g(c_2, T)\frac{\partial T}{\partial c_2} - \Delta \frac{\partial T}{\partial a} + \cdots. \end{aligned}$$

Hence,

$$h(c_2, T) = g(c_2, T)\frac{\partial T}{\partial c_2} + \frac{\partial T}{\partial a}, \qquad (14.22.3)$$

with the initial condition

$$T(c_1, c_2, 0) = c_1. \qquad (14.22.4)$$

14.23. Branching Processes

Let us now turn to the third type of stochastic iteration process. Consider the following. An object born at time $t = 0$ has a random lifelength with probability distribution $G(t)$. At the end of its life it is replaced by a random number of similar objects, with p_r being the probability that the new number of objects is r, $r = 0, 1, 2, \ldots$. These probabilities are taken to be independent of absolute time, of the age of the object when it dies, and of the number of objects at the time.

Let $z(t)$ be the number of objects present at time t. This is a random function which we call an age-dependent branching process. Let

$$f_r(t) = \text{Prob}(z(t) = r), \qquad r = 0, 1, \ldots, \qquad (14.23.1)$$

and set

$$f(s, t) = \sum_{r=0}^{\infty} f_r(t)s^r = E(s^{z(t)}). \qquad (14.23.2)$$

Then, setting

$$h(s) = \sum_{r=0}^{\infty} p_r s^r, \qquad (14.23.3)$$

we have the nonlinear integral equation

$$f(s, t) = s(1 - G(t)) + \int_0^t h(f(s, t - t_1))\, dG(t_1), \qquad |s| \leq 1, \quad t \geq 0. \qquad (14.23.4)$$

Exercises

1. Derive (14.23.4).
2. Show that if dG is a delta-function, this branching process reduces to iteration.
3. Consider the case where $dG = \lambda e^{-\lambda t}$, $\lambda > 0$, or a sum of exponential terms.
4. Show that for each s, $0 \leqslant s \leqslant 1$, there is one nonnegative solution of (14.23.3).
5. Obtain a linear integral equation of renewal type for $v(t) = E(z(t))$ in two ways, by differentiation of (14.23.3) and directly, namely

$$v(t) = 1 - G(t) + m \int_0^t v(t - t_1)\, dG(t_1),$$

where $m = h'(1)$.

6. Solve this equation using the Laplace transform and obtain asymptotic behavior under various assumptions using the contour integral representation of the Laplace inverse: see Chapter 16. For the foregoing see

R. Bellman and T. E. Harris, "On the Theory of Age-Dependent Stochastic Branching Processes," *Proc. Nat. Acad. Sci.*, Vol. 34, 1948, pp. 601–602

R. Bellman and T. E. Harris, "On Age-Dependent Binary Branching Processes," *Annals Math.*, Vol. 55, 1952, pp. 280–295.

N. Levinson, "Limit Theorems for Age-Dependent Branching Processes," *Illinois J. Math.*, (Vol. 4, 1960, pp. 110–118.)

14.24. Closure of Operations

In the conclusion of this chapter, let us put our results into broader context. What we have been studying is a part of a general pattern that is slowly emerging, a general theory of closure of operations.

It began with the invention of the logarithmic operation by Napier and Briggs. Their fundamental contribution was the observation that multiplication could be converted to addition with the aid of a transcendental operation, the taking of a logarithm. Thus,

$$\log(ab) = \log a + \log b. \tag{14.24.1}$$

In the foregoing pages we have shown that certain other classes of transformations could be reduced to multiplications plus a few additional fixed transformations.

There are many binary operations in analysis, the most important of which is the convolution of two functions

$$f * g = \int_0^t f(t_1)g(t - t_1)\, dt_1 = \int_0^t f(t - t_1)g(t_1)\, dt_1 . \qquad (14.24.2)$$

In solving the renewal equation, we have made use of the fact that the Laplace transform converts this operation into multiplication, namely

$$L(f * g) = L(f)\, L(g) \qquad (14.24.3)$$

where

$$L(f) = \int_0^\infty f(t)e^{-st}\, dt. \qquad (14.24.4)$$

This result is one of the foundations of linear analysis.

Many different "convolutions" of two functions can be introduced. Thus, we can introduce the "maximum transform," with the convolution

$$f \circ g = \max_{0 \leqslant t_1 \leqslant t} [f(t_1)\, g(t - t_1)] \qquad (14.24.5)$$

and show that under appropriate conditions

$$M(f \circ g) = M(f)\, M(g) \qquad (14.24.6)$$

where

$$M(f) = \max_{t \geqslant 0} [e^{-st} f(t)]. \qquad (14.24.7)$$

This result plays an important role in certain classes of dynamic programming and variational processes and in the general theory of convexity as we have discussed in Chapter 12. It is closely related to the Legendre–Fenchel transform discussed in the chapter on dynamic programming. It bears the same correspondence to this that the Laplace transform bears to the Fourier transform.

14.25. Number of Arithmetic Operations

A basic consideration in developing algorithms for use by a digital computer is the number of arithmetic operations involved. We are specifically interested in the number of multiplications since they consume about ten times as much time as additions at the present.

Consequently, we are concerned with various ways of scheduling the component parts of algorithms so as to minimize the number of multiplications. Although these appear to be formidable problems, recently some important results have been obtained. Let us briefly discuss some of them.

Consider first the multiplication of two matrices

$$C = AB, \qquad c_{ij} = \sum_{k=1}^{N} a_{ik} b_{kj}, \tag{14.25.1}$$

A straightforward approach requires N^3 multiplications and $(N - 1)N^2$ additions. It was widely believed for quite some time that nothing better could be achieved. Then Winograd showed that half the multiplications could be traded for additions, using some auxiliary functions. Using some more sophisticated techniques, Strassen showed that only $N^{2.8}$ multiplications were required.

Consider next the problem of the evaluation of 2^N. A straightforward approach requires $N - 1$ multiplications. Using successive squaring, $2^2 = 4$, $4^2 = 16, \ldots$, this can be reduced to approximately $2 \log_2 N$. The optimal procedure remains unknown. The corresponding problem for $2^N 3^M$ seems even more difficult.

Consider finally the evaluation of a polynomial, say $a_0 n^3 + a_1 n^2 + a_2 n + a_3$. Although direct approach requires five multiplications, the classical procedure of Horner,

$$((a_0 n + a_1)n + a_2)n + a_3 \tag{14.25.2}$$

requires only three multiplications. Ostrowski showed that this procedure of Horner was the best possible for polynomials of degree four or less, while, more recently, Pan showed that the result was valid for polynomials of any degree.

Nonetheless, this does not end the matter since interesting questions remain for the case when the same polynomial is to be evaluated a large number of times and for polynomials of several variables.

Miscellaneous Exercises

1. For the case where

$$u' = -u + g_1(u, v), \qquad u(0) = c_1,$$

$$v' = -2v + g_2(u, v), \qquad v(0) = c_2,$$

for both types of possible extension of relative invariants given in Sec. 14.19, show how to determine the coefficients recurrently and discuss the convergence of the series thus obtained.

2. If $u_{n+1} = au_n + g(u_n) + r$, $u_0 = c$, obtain possible forms of asymptotic behavior of u_n, where $g(u)$ is analytic for $|u| \ll 1$, lacking constant and first-order terms, and $|c| \ll 1$, $0 < |a| < 1$. Under what conditions on the sequence $\{r_n\}$ can we assert that

$$\phi(c) = \lim_{n\to\infty} \frac{u_n}{a^n}$$

exists?

3. Consider next the scalar recurrence relation

$$u_{n+1} = au_n + g(u_n) + r_n, \qquad u_0 = c.$$

4. Let, for $k = 1, 2, \ldots$,

$$\phi_k(c) = \lim_{n\to\infty} \frac{u_n}{a^n},$$

where the recurrence relation holds for $n \geqslant k$, with $u_k = c$. Show that

$$a\phi_k(c) = \phi_{k+1}(ac + g(c) + r_k).$$

5. What is the situation when r is a random variable? Assume that the r_n are independent with the same distribution function. What do we mean by asymptotic behavior?

6. Consider a functional equation of the type arising in dynamic programming,

$$f_n(c) = \max_y f_{n-1}(T(c, y)), \qquad f_0(c) = c,$$

where

$$T(c, y) = a_1(y)c + a_2(y)c^2 + \cdots,$$

for $|c| \ll 1$ and all y, and

$$0 < a_0 \leqslant a_1(y) \leqslant b_0 < 1$$

for all y. If $b = \max_y a_1(y)$, can we assert that $\varphi(c) = \lim_{n\to\infty} f_n(c)/b^n$ exists for $|c| \ll 1$? Is it true that $b\varphi(c) = \max_y \varphi(T(c, y))$?

7. If $a_{n+1} = a_n(1 - a_n^2)$, $0 < a_1 < 1$, show that $a_n \to 0$ as $n \to \infty$, and more precisely that as $n \to \infty$

$$a_n = \frac{1}{\sqrt{2n}} - \frac{3 \log n}{8\sqrt{2}n^{3/2}} + O(n^{-3/2})$$

S. Spital-Yo Chang, *Amer. Math. Monthly*, Vol. 74, 1967, p. 334.

8. If $z_{n+1} = z_n - a z_n^{p+1}$, $a > 0$, and if $0 < | z_1 | < b$, $| \arg z_1 | < \pi/(p - \epsilon)$, then $z_n = [npa + O(n^{1-1/p})]^{-1/p}$ as $n \to \infty$.

L. Fatou, "Sur les equations fonctionelles," *Bull. Soc. Math. France*, Vol. 47, 1919, pp. 161–271.

9. Consider the transformation

$$f(z) = \lambda z + \sum_{k=2}^{\infty} a_k z^k,$$

convergent for $| z | < b$. If $\lambda^n = 1$ and the transformation is stable, then $f^{(n)}(z) = z$.

10. If $\lambda = e^{2\pi i a}$, a real and irrational, the Abel–Schroder equation possesses exactly one formal solution. See

H. Russman, "On the Iteration of Analytic Functions," *J. Math. Mech.*, Vol. 17, 1967, pp. 523–532,

for many additional results.

11. If $| \lambda | \geqslant 1$, it is necessary and sufficient for stability, in the sense that the transformation $z_{n+1} = f(z_n)$ is such that $| z_0 | \leqslant r_0$ implies $| z_n | \leqslant r$, that $| \lambda | = 1$, and that the equation $\varphi(\lambda z) = f(\varphi(z))$ possesses a convergent solution.

C. L. Siegel, *Vorlesungen uber Himmelsmechanik*, Springer, New York, 1954.

12. Consider the functional differential equation

$$u'(t) = g(u(t), u(h(t))).$$

Show that by introduction of a function p satisfying $h(p(t)) = p(t - b)$, we can reduce this to a differential-difference equation. (The result is due to Laplace. See

R. Bellman and K. L. Cooke, "On the Computational Solution of a Class of Functional Differential Equations," *J. Math. Anal. Appl.*, Vol. 12, 1965, pp. 495–500.)

13. Consider the application of relative invariants to the solution of two-point boundary-value problems of the form $u' = g(u, v)$, $u(0) = c_1$, $v' = h(u, v)$, $v(a) = c_2$. Let φ and ψ be such that $\varphi(u, v)' = \lambda_1\varphi(u, v)$, $\psi(u, v)' = \lambda_2\psi(u, v)$. Let u, v be replaced by the variables w and z, where $\varphi(u, v) = w$, $\psi(u, v) = z$, and solve the corresponding two-point boundary-value problem for w and z. What is required to make the procedure rigorous?

14. Consider the differential equation

$$u' = g(u) + h(t), \qquad u(a) = c,$$

where

$$g(u) = -u + \cdots, \qquad \text{for} \quad |u| \ll 1.$$

Let

$$\phi(a, c) = \lim_{t\to\infty} e^t u(t)$$

for $|c| \ll 1$ and suitable conditions on $h(t)$. Show that

$$-\phi = \frac{\partial\phi}{\partial a} + (h(a) + g(c))\frac{\partial\phi}{\partial c}.$$

Obtain an expansion for $\varphi(a, c)$ in powers of c. Determine the nature of $\varphi(a, c)$ when $h(t) = e^{-bt}$, $b > 0$, and when $h(t) = t^{-b}$, $b > 0$.

15. Let $h(u) = a_0 + a_1u + \cdots + a_nu^n + \cdots$. Write

$$h(u)^j = \sum_{i=0}^{\infty} a_{ij}u^i.$$

Evaluate the determinant $|a_{ij}|$, $i = 0,\ldots, n-1$, $j = 1,\ldots, n$.

16. Let $h(u) = a_1u + \cdots + a_nu^n + \cdots$. Write

$$h^{(j)}(u) = \sum_{i=1}^{\infty} b_{ij}u^i.$$

Evaluate the determinant $|b_{ij}|$, $i, j = 1,\ldots, n$.

17. Let $\{a_n\}$ be a sequence with the property that $0 < a_n < a_{n+1} < a_{f(n)}$ for all n. Then $\sum a_n$ diverges if $f(n) = n(n+1)$. (Graham and Lewett.)

18. The series diverges if $f(n) = n^2$ (D. J. Newman). See
K. A. Post, "A Combinatorial Lemma Involving a Divergence Criterion for Series of Positive Terms," *Amer. Math. Mech.*, Vol. 77, 1970, pp. 1085–1087.)

19. Let $g(t, c)$ be an entire analytic function of t dependent on a vector parameter c, and write $f_k(t) = g(t, b_k)$, where b_k is a particular choice of c. Consider the function $h_N(t) = f_1(f_2 \ldots(f_N(t)\ldots.)$ Can we choose the b_k so that $h_N(t)$ is an arbitrarily good approximation to a given entire analytic function $\varphi(t)$ in the sense of agreement of $\psi(N)$ terms in the power series where $\psi(N) \to \infty$ as $N \to \infty$, or any other convenient norm?

Clearly, this cannot be done for arbitrary $g(t, c)$, e.g., $g(t, c) = c_0 + c_1 t$. How do we recognize the class of functions for which this is possible? Is it possible for $g(t, c) = c_0 + c_1 t + c_2 t^2$ or $g(t, c) = c_1 e^{c_2 t}$, and, if so, what degree of approximation can be attained?

20. Consider the equation $Ax = t$, where A is a positive definite linear operator with spectrum contained in the interval $[a, b]$. Consider the successive approximation determined by $x_{n+1} = x_n + \alpha_n(Ax_n - f)$, when the α_n are random variables. Can we choose the α_n so that there is convergence with probability one? See
J. V. Vorobev, "A Random Iteration Process II" (Russian), *Z. Vycisl. Mat. i Mat. Fiz.*, Vol. 5, 1965, pp. 787–795; *Math. Rev.*, Vol. 34, 1967, No. 5, 7004.

21. Consider the sequence $u_{n+1} \equiv au_n(p)$, $u_0 \equiv b$ (modulo a prime p) where a and b are in the same residue class. Show that $u_{n+p} \equiv u_n(p)$ for all n (Fermat's theorem).

22. Consider the sequence $u_{n+1} \equiv (a_1 u_n + a_2 u_n{}^2)(p)$, $u_0 \equiv b$. Do corresponding results hold?

23. Consider the case $X_{n+1} \equiv AX_n(p)$, $X_0 = B$. Do corresponding results hold?

24. Consider the case $u_{mn} \equiv au_{m-1,n} + bu_{m,n-1}(p)$, u_{m0} and u_{0n} specified. Do corresponding results hold?

25. Let u, v, and w be three functions defined for $t \geqslant 0$ and connected by the convolution relation $w(t) = \int_0^t u(s)v(t - s)\, ds$. Suppose that u, v, w are known on subsets of $[0, \infty]$, S_1, S_2, and S_3, respectively. What assumptions concerning S_1, S_2, and S_3 and the behavior of u, v, and w suffice to determine u, v, and w for $t \geqslant 0$?

26. Consider the same problem for the case where

$$\text{(a)} \quad w_n = \sum_{k=0}^{n} u_k v_{n-k}, \qquad n = 0, 1, \ldots,$$

$$\text{(b)} \quad w_n = \sum_{k|n} u_k v_{n/k}, \qquad n = 1, 2, \ldots$$

($k \mid n$ means that k divides n).

27. What algorithms exist for obtaining u, v, and w?

28. Let A and B be two positive definite matrices and let X be a positive definite matrix such that $X \geqslant A$ and $X \geqslant B$. Denote the set of all such X by the expression max(A, B). Show that max(A, B) is convex.

    Find the elements X in max(A, B) which minimize tr X and tr(X^2).

    Similarly consider max(A, B, C). Is it true that max (A, B, C) = max(max(A, B), C) = max (A, max(B, C)) where max(max(A, B), C) and max(A, B, C) are defined in obvious fashion?

29. Let X_1 and X_2 be two complex $N \times N$ matrices which do not commute. Introduce the norm of an $N \times N$ matrix Z by $\| Z \|^2 = \operatorname{tr}(Z\bar{Z}')$. What is the minimum of $\| X_1 - Y_1 \|^2 + X_2 - Y_2 \|^2$ over all $N \times N$ matrices Y_1 and Y_2 which do commute? Is there a general inequality connecting $\| X_1 - Y_1 \|^2$, $\| X_2 - Y_2 \|^2$, $\| X_1X_2 - X_2X_1 \|$, $\| Y_1Y_2 - Y_2Y_1 \|$?

30. Let X_1 and X_2, as before, not commute. What is the minimum value over all complex scalars a_i of

$$\| X_2 - \sum_{k=0}^{N-1} a_k X_1{}^k \|^2,$$

where N is the dimension of X?

31. The study of the firing of nerves leads to equations of the following type:

$$\frac{du_1}{dt} = a_{11}u_1 + a_{12}u_2, \qquad u_1(0) = c_1,$$

$$\frac{du_2}{dt} = a_{21}u_1 + a_{22}u_2, \qquad u_2(0) = c_2,$$

for $0 \leqslant u_1, u_2 < 1$, where $0 \leqslant c_1, c_2 < 1$, $a_{ij} \geqslant 0$. As soon as either u_1 or u_2 attains the value 1, it instantaneously returns to zero value, leaving the other value unchanged, and the above equations take over again.

What are the periodic and ergodic properties of solutions of equations of this type? Are there simple analytic expressions for the solution at time t?

32. If $u_n \geqslant 0$ and $u_{m+n} \leqslant u_m + u_n$, $m, n \geqslant 0$, then u_n/n approaches a limit as $n \to \infty$ (Fekete–Polya–Szego).

33. Consider the expression $u_n(p) = \sum_{k=0}^{n-1} f(T^k p)$, where $p \in R$ and $Tp \in R$. Write $v_n = \max_p u_n(p)$. Show that $v_{n+m} \leqslant v_n + v_m$ and thus that $v_n/n \to r$ as $n \to \infty$.

34. How can one go from this to the existence of a limit for $u_n(p)/p$? See

    R. Bellman, "Functional Equations in the Theory of Dynamic Programming—XI: Limit Theorems," *Rend. Circ. Mat. Palermo*, Vol. 8, 1959, pp. 1–3.

35. Let the one-dimensional sequence $\{x_n\}$ be generated by means of the recurrence relation

$$x_{n+1} = ax_n + g(x_n) + r_n , \qquad x_0 = c,$$

where a and c are real, $\{r_n\}$ is a sequence of independent random variables with known identical distribution functions, and $g(x)$ is a feedback control function defined as follows:

$$\begin{aligned} g(x) &= 0, & |x| &< b, \\ &= k, & x &< -b, \\ &= -k, & x &> b, \end{aligned}$$

where $k > 0$.

What possiblities exist for the limiting distributions of x_n, suitably normalized?

Consider also the special case where $r_n = r$, a constant.

36. Use the functional equation $f(x) = (1 + xq)f(qz)$ to derive the identity

$$f(x) = \sum_{n=0}^{\infty} (1 + xq^n) = 1 + \sum_{n=1}^{\infty} \frac{x^n q^{n(n+1)/2}}{(1-q)(1-q^2)\cdots(1-q^n)} .$$

37. Obtain the result by using an appropriate partial fraction expansion. See

    R. Bellman, "The Expansion of Some Infinite Products," *Duke Math. J.*, Vol. 24, 1957, pp. 353–356,

    Exercise 30, pp. 46–47, Volume I, and

    R. Bellman, *A Brief Introduction to Theta Functions*, Holt, New York, 1961.)

38. Consider the equation $u'' + a(t)u = 0$, $u(0) = c$, $u(T) = d$. Guess an initial value $u'(0) = c_1$ and determine the solution of the initial value problem $v'' + a(t)v = 0$, $v(0) = c$, $v'(0) = c_1$ and thus $v'(T)$; call this d_1. Consider the new initial value problem $w'' + a(t)w = 0$, $w(T) = d$, $w'(T) = d_1$. Determine the corresponding value $w'(0)$, and so on. When does this procedure converge to the solution of the original problem?

39. Consider the corresponding vector equation,

$$x'' + A(t)x = 0, \qquad x(0) = c, \qquad x(T) = d.$$

40. Can the procedure be modified to converge in all cases where the original equation has a unique solution? See

    R. Bellman, "On the Iterative Solution of Two-Point Boundary Value Problems," *Boll. U.M.I.*, Vol. 16, 1963, pp. 145–149.

    See also Exercise 15 at the end of Chapter 16.)

Bibliography and Comments

§14.1. For a discussion of the connections between the theory of iteration and celestial mechanics with references to the work of Poincaré, Birkhoff and others, see

C. L. Siegel, *Vorlesungen uber Himmelsmechanik*, Springer, New York, 1954.

See also

J. Hadamard, "Two Works on Iteration and Related Questions," *Bull. Amer. Math. Soc.*, Vol. 50, 1944, pp. 67–75.

G. I. Targonski, *Seminar on Functional Operators and Equations*, *Lecture Notes in Math.*, Vol. 33, Springer–Verlag, Berlin, New York, 1967.

S. Diliberto, "A New Technique for Proving the Existence of Analytic Functions, Ordinary Differential Equation," 1971 *NRL-MRC Conference*, Academic Press, New York, 1972, pp. 71–82.

For a discussion of the many ways in which iteration enters into modern science, see

T. E. Harris, *The Theory of Branching Processes*, Springer–Verlag, Berlin, 1963.

C. Mode, *Multitype Branching Processes*, Elsevier, Amsterdam, 1971.

§14.5. See

G. Koenigs, "Recherche sur les Integrales de Certaines Equations Fonctionelles," *Annales Sci. de l'École Normale Superieure* (3), Vol. 1, 1884, Supplement, pp. 3–41.

G. Koenigs, "Nouvelles Recherches sur les Equations Fonctionelles," *Annales Sci. de l'École Normale Superieure* (3), Vol. 2, 1885, pp. 385–404.

For detailed discussions of the equations of Schroder, see

M. Urabe, "Equations of Schroder," *J. Sci. Hiroshima Univ.*, Ser. A, Vol. 15, 1951, pp. 113–131.

M. Urabe, "Equations of Schroder," *J. Sci. Hiroshima Univ.*, Ser. A, Vol. 15, 1951, pp. 203–233.

M. Urabe, "Invariant Varieties for Finite Transformations," *J. Sci. Hiroshima Univ.*, Ser. A, Vol. 16, 1952, pp. 47–55.

M. Urabe, "Iteration of Certain Finite Transformations," *J. Sci. Hiroshima Univ.*, Ser. A, Vol. 16, 1952, pp. 471–486.

M. Urabe, "Application of Majorized Group of Transformations to Functional Equations," *J. Sci. Hiroshima Univ.*, Ser. A, Vol. 16, 1952, pp. 267–283.

§14.9. See

R. Bellman, "Stochastic Transformations and Functional Equations," *IRE Trans.*, Vol. AC–7, 1962, p. 120.

§14.10. There are many important and difficult questions here, particularly in connection with asymptotic behavior. See

R. Bellman, "Limit Theorems for Noncommutative Operations—I," *Duke Math. J.*, Vol. 21, 1954, pp. 491–500.

H. Furstenberg and H. Kesten, "Products of Random Matrices," *Amer. Math. Stat.*, Vol. 3, 1960, pp. 457–469.

and the references in Volume I.

Because of the phenomenon of round-off error, every numerical process for the solution of differential equations, and more generally, any iterative process such as that for example involved in the use of the Newton–Raphson–Kantorovich technique or quasilinearization, leads to a stochastic iteration process.

For a discussion of these matters and of the interesting phenomenon of oscillatory numerical convergence, see

M. Urabe, "Error Estimation in Numerical Solution of Equations by Iteration Processes," *J. Sci. Hiroshima Univ.*, Ser. A, Vol. 26, 1962, pp. 77–91.

§14.11. This was indicated in

R. Bellman and J. M. Richardson, "Relative Invariants, Closure and Stochastic Differential Equations," *J. Math. Anal. Appl.*, Vol. 14, 1966, pp. 294–296.

§14.15. See

A. Grévy, "Étude sur les Équations Fonctionelles," *Annales de Sci. l'École Normali* (3), Vol. 11, 1894, pp. 249–323.

L. Leau, "Étude sur les Équations Fonctionelles," *Annales de Sci. l'Ecole Normali* (3), Vol. 11, 1894, pp. 249–323.

L. Leau, "Étude sur les Équations Fonctionelles à une ou à Plusieurs Variables," *Annales de la Faculté des Sci. de Toulouse*, Vol. 11, 1897, pp. 1–110.

R. Bellman, "The Iteration of Power Series in Two Variables," *Duke Math. J.*, Vol. 19, 1952, pp. 339–347.

H. Topfer, "Komplexe Iterations Indizes Ganzer und Rationales Funktionen," *Math. Ann.*, Vol. 121, 1949–1950, pp. 191–222.
G. F. Schubert, "Solution of Generalized Schroeder Equation in Two Variables," *J. Austral. Math. Soc.*, Vol. 4, 1964, pp. 410–417.

§14.19. See

R. Bellman, "Relative Invariants in the Commensurable Case," *J. Math. Anal. Appl.*, Vol. 28, 1969, pp. 400–404,

and the Schubert paper cited above for a detailed analysis.

§14.20. See

R. Bellman, "Two-Point Boundary-Value Problems and Iteration, *Aequationes Math.*, Vol. 2, 1969, pp. 167–170.

§14.23. See the book by T. Harris cited above and

D. G. Kendall, "Branching Processes Since 1873," *J. London Math. Soc.*, Vol. 41, 1966, pp. 385–406.

§14.30. See

S. Winograd, "The Number of Multiplications Involved in Computing Certain Functions," *Proc. IFIP Congr.*, 1968.
V. Strassen, *Z. Angew. Math.* (to appear).
H. Kato, *On Addition Chains*, Ph. D. Thesis, Univ. of Southern California, Dept. of Math., 1969.
V. Ja Pan, "On Means of Calculating Values of Polynomials," *Math. Rev.* No. 6994, Vol. 34, 1967.
A. Ostrowski, *On Two Problems in Abstract Algebra Connected with Horner's Rule, Studies in Mathematics and Mechanics Presented to Richard von Mises*, Academic Press, New York, 1954, pp. 40–48.
F. J. Smith, "An Algorithm for Summing Orthogonal Polynomial Series and Their Derivatives with Application to Curve-Fitting and Interpolation," *Math. Comp.*, Vol. 19, 1963, pp. 33–36.

Chapter 15 INFINITE SYSTEMS OF ORDINARY DIFFERENTIAL EQUATIONS AND TRUNCATION

15.1. Introduction

In this chapter we wish to touch upon some topics connected with infinite systems of ordinary differential equations, an area relatively unexplored despite the fact that it is interesting and important in both theory and application. Our objective is to show, in the text and the exercises, some of the numerous ways in which equations of this nature arise and some of the many questions encountered in obtaining analytic and numerical results.

The idea of reducing problems involving functional equations to that of solving infinite sets of linear and nonlinear equations is an old one associated with the very beginnings of modern analysis. Contributions were made by mathematicians such as Fourier, Hill, Lichtenstein, F. Riesz and Poincaré. The feasibility, however, of the approach has changed dramatically as a consequence of the development of the digital computer.

To use the digital computer to study a number of the important equations of analysis, a systematic closure of operations is required. Formulations of equations in terms of derivatives and integrals must be supplemented by associated equations involving algorithms requiring only a finite number of arithmetic operations. In particular, infinite systems must be replaced by finite ones. A discussion of such matters will be begun in this chapter and continued in the next.

We start with a brief review of methods for dealing with ordinary differential equations and continue to a sketch of some of the standard ways in which discretization is carried out for partial differential equations. We shall focus upon the particular equation,

$$u_t = u_{xx} + u^2, \qquad u(x, 0) = g(x), \quad 0 \leqslant x \leqslant 1, \tag{15.1.1}$$

with $g(x)$ periodic of period one. Then we shall describe some alternate approaches which lead to infinite systems of ordinary differential equations. From this we are led naturally to problems of truncation. In particular, we shall discuss a venerable procedure known as the "methode des reduites."

A point worth stressing is that it is quite advantageous to possess a large number of different approaches to the study of partial differential and more general functional equations. Some are particularly useful for establishing existence and uniqueness of solution, some are more suited for computational purposes and some are well qualified to verify the presence of certain important properties such as nonnegativity. Essentially, each equation needs a set of individualized approaches.

15.2. Ordinary Differential Equations and Difference Methods

As pointed out above, to obtain a computational solution of a system of differential equations of the form

$$\frac{dx}{dt} = g(x), \qquad x(0) = c, \tag{15.2.1}$$

by means of a digital computer, we must employ some arithmetic algorithm. One standard way of accomplishing this is to use the approximation

$$\frac{dx}{dt} \cong \frac{x(t + \Delta) - x(t)}{\Delta}, \tag{15.2.2}$$

where $\Delta > 0$ and replace (15.2.1) by the difference equation

$$\frac{y(t + \Delta) - y(t)}{\Delta} = g(y(t)), \qquad y(0) = c, \tag{15.2.3}$$

where t now assumes only the discrete values $t = 0, \Delta, 2\Delta, \ldots$.

Obtaining an approximate analytic or numerical solution to the original equation has thus been reduced to the task of iteration of the transformation

$$T(y) = y(t) + \Delta g(y(t)), \qquad y(0) = c, \tag{15.2.4}$$

where $T^n y = y(n\Delta)$.

If the components of g can be calculated arithmetically in terms of the components of y, certainly the case if the components are polynomials in the components of y, the calculation can readily be carried out. Otherwise, some further approximations are required.

15.3. Stability Considerations

Stability enters in a number of ways as a consequence of the fact that there are three sources of error. The first is a result of the use of a discretization technique such as (15.2.2), a second lies in the evaluation of $g(y)$ and a third resides in the use of the arithmetic procedures of the computer in general. Let us discuss these in turn, pausing to reiterate the point that computing must be regarded as a control process of deterministic, stochastic and adaptive type, both "open loop" and "closed loop" aspects.

Let us consider the difference equation of (15.2.3). Clearly, a question of paramount interest is to be able to determine when $y(t) \equiv y(t, \Delta)$ furnishes a satisfactory approximation to the value $x(t)$.

Observe that there are actually two distinct classes of problems here. The first is that of establishing this equivalence in the case where it is known in advance that the differential equation possesses a solution with various properties. The second is that of establishing ab initio that the function $y(t, \Delta)$ possesses a limit as $\Delta \to 0$ and that this limit is the desired solution $x(t)$. In many cases where the discrete problem has immediate physical significance this latter is the more meaningful question. Indeed, one can, if one wishes, bypass the continuous process, defining it if desired solely in terms of the limit of the discrete version.

As we have previously indicated, this second problem may be regarded as a stability problem associated not with the original differential equation, but rather with the difference equation. Thus, if we write, in some interval $[0, t_0]$,

$$\begin{aligned} x(t + \Delta) &= x(t) + \Delta x'(t) + \frac{\Delta^2}{2} x''(t) + \cdots, \\ \frac{x(t + \Delta) - x(t)}{\Delta} &= x'(t) + O(\Delta), \end{aligned} \tag{15.3.1}$$

the differential equation becomes

$$\frac{x(t + \Delta) - x(t)}{\Delta} = g(x(t)) + O(\Delta), \tag{15.3.2}$$

as compared with the equation of (15.2.3).

15.4. Round-off Error

Digital computer arithmetic is not exact. It is necessarily approximate arithmetic to a certain number of significant figures with the number of significant figures dependent upon the time and effort that one wishes to expend. Introducing the error $r(t)$ caused by the various arithmetic operations, (15.2.3) takes the form

$$z(t + \Delta) = z(t) + \Delta g(z(t)) + r(t). \tag{15.4.1}$$

The function $r(t)$ should really be written $r(z, t)$, a very complicated function of z. In the absence of any precise information concerning its true nature, it is most conveniently regarded as a random vector, the usual scientific device to cope with complexity. Questions arise immediately concerning the relation between $z(t)$ and $y(t)$, and thus between $z(t)$ and $x(t)$. This leads into the important area of difference and differential equations with stochastic terms and is an important motivation for the study of equations of this nature.

One way of diminishing the effect of the error term $r(t)$ is to use a more sophisticated approximation than that appearing in (15.2.2). For example, we might write

$$\frac{dx}{dt} \cong \frac{x(t + \Delta) - x(t - \Delta)}{2\Delta} \tag{15.4.2}$$

with an error term which is $O(\Delta^2)$ instead of $O(\Delta)$ as in (15.2.2). One disadvantage of this procedure is that now both $x(t - \Delta)$ and $x(t)$ must be stored to obtain $x(t + \Delta)$. More immediate disadvantages of this approximation are connected with further stability considerations. Thus, for example, if $g(x) = Ax$, the difference equation

$$\frac{y(t + \Delta) - y(t - \Delta)}{2\Delta} = Ay(t) \tag{15.4.3}$$

can possess disturbing extraneous solutions which combined with round-off error can rapidly overwhelm the desired solution.

Many more sophisticated kinds of difference approximations are now available. It is quite an art to chose an appropriate balance between possible error in the final result, and excessive time required for calculation.

Although these control aspects of computing are negligible for low dimensional systems, they become of paramount importance when the dimension of x is large. At the present time we can think comfortably of solving systems of dimension one hundred. Within a few years, particularly with parallel and special purpose computers, it will be commonplace to treat systems of dimension one thousand and ten thousand. Within ten years, we will be able to handle one hundred thousand simultaneous differential equations. This capability strongly affects our approach to partial differential equations.

The difficulties involved in storing the instructions for calculating the components of $g(x)$ are not serious in the important case where the differential equations are obtained from some underlying functional equation of more complex type such as a partial differential equation. As usual, we will, however, be forced to pay for this in time. We will not delve into any detailed considerations here.

Let us merely point out that the task of solving large systems of ordinary differential equations is an important one which requires a good deal of effort and ingenuity. In particular, we need methods which yield the desired information without the simultaneous calculation of a good deal of extraneous data, a theory of selective calculation. This will be further discussed in the next chapter.

15.5. Linear Partial Differential Equations

Let us now consider the question of numerical solution of a linear partial differential equation, say

$$\begin{aligned} k(x)\, u_t &= u_{xx}\,, \qquad t > 0, \quad 0 < x < 1 \\ u(x, 0) &= g(x), \qquad 0 < x < 1, \end{aligned} \tag{15.5.1}$$

with $k(x)$, $g(x)$ and the solution $u(x, t)$ periodic of period one.

Employing finite difference approximations of the type considered above for ordinary differential equations, we obtain the associated finite difference equation

$$k(x) \left[\frac{v(x, t + \Delta) - v(x, t - \Delta)}{2\Delta}\right] = \left[\frac{v(x + \delta, t) + v(x - \delta, t) - 2v(x, t)}{\delta^2}\right] \tag{15.5.2}$$

where Δ and δ are positive quantities. The values of t are restricted to $0, \Delta, 2\Delta,\ldots$, and the values of x to $0, \delta, 2\delta, 3\delta,\ldots$. The values at $t = 0$ and at $x = 0$, $x = 1 = N\delta$ are determined by the initial and boundary conditions. Use of (15.5.2) then permits the recurrent calculation of $v(x, t)$ for $t = \Delta, 2\Delta,\ldots$ and $x = \delta, 2\delta,\ldots, (N - 1)\delta$, given the values at $t = 0$ and $x = 0$ and 1. The $(N - 1)$-dimensional vector $v(x, t)$, for $t = M\delta$, $x = \delta, 2\delta,\ldots, (N - 1)\delta$ is calculated using the corresponding vector for $t = (M - 1)\delta$.

This is a closure technique. The original transcendental problem of solving a partial differential equation has been reduced to an arithmetic algorithm involving the iteration of a linear transformation. The question arises as to the connection between the solution of the linear difference equation of (15.5.2), $v(x, t)$, and the solution of the partial differential equation of (15.5.1) as well as the connection between $v(x, t)$ and the numerical results obtained when (15.5.2) is used to calculate $v(x, t)$ with the aid of a digital computer. As pointed out above, the digital computer may be regarded as producing a solution of an associated inhomogeneous equation where the forcing term is a consequence of various errors. These questions can be regarded as stability problems associated with the difference equation of (15.5.2).

It turns out that there are severe restrictions on the use of (15.5.2). In order to ensure numerical stability we must have $\Delta \cong \delta^2$. We leave the derivation of this classical result of Courant, Friedrichs and Loewy as an exercise. Since the step size in space must in general be taken small to ensure accuracy, this means that the step size in time must be taken quite small (e.g., $\delta = 0.01$ implies that $\Delta = 0.0001$), and thus that the calculation will be extremely time-consuming. In addition, the end result is subject to an accumulation of round-off errors.

One way to overcome the restriction $\Delta \cong \delta^2$, an annoying feature of a recurrence relation of the foregoing type, is to replace (15.5.2) by a slightly modified version, say an implicit recurrence relation such as

$$k(x)\left[\frac{u(x, t+\Delta) - u(x, t-\Delta)}{2\Delta}\right] = \left[\frac{u(x+\delta, t+\Delta) + u(x-\delta, t+\Delta) - 2u(x, t+\Delta)}{\delta^2}\right]. \tag{15.5.3}$$

Observe that both the left-hand and right-hand side contains the timetag $t + \Delta$. This procedure, introduced by Von Neumann, possesses the great merit of being unconditionally stable for any Δ, $\delta > 0$. Again, we leave this as an exercise. In return for this, however, one faces the onerous task of solving a large system of simultaneous algebraic equations to calculate $u(x, t + \Delta)$, $x = \delta, 2\delta, \ldots$, given the values of $u(x, t - \Delta)$ for $x = \delta, 2\delta, \ldots$.

Operations of this nature have been raised to a highly developed art over the last twenty years and analyzed in careful detail. Nonetheless, it is not routine to carry over the methods developed for linear equations to the case where the equation is nonlinear, say

$$k(x)\, u_t = u_{xx} + u^2 \tag{15.5.4}$$

or

$$u_t = u u_x , \tag{15.5.5}$$

nor even to linear equations with variable coefficients over irregular regions. Indeed, there is no satisfactory stability theory for nonlinear partial difference equations, no more than there is for nonlinear differential or difference equations.

Nonlinear equations are assuming roles of greater and greater importance as more realistic models of scientific phenomena corresponding to higher energies are studied with greater frequency and as control theory is assuming a greater role. Hence, it is essential to have a number of powerful computational algorithms available. One may work when others fail.

15.6. Partial Discretization

One simple way to bypass the restriction on Δ is to take it equal to zero. By this we mean that we compromise between (15.5.1) and (15.5.2), and use an equation such as

$$k(x)\frac{\partial u(x, t)}{\partial t} = \left(\frac{(u(x+\delta, t) + u(x-\delta, t) - 2u(x, t)}{\delta^2}\right), \tag{15.6.1}$$

Here $t \geqslant 0$ and x once again assumes the values $\delta, 2\delta, \ldots, (N-1)\delta$. Writing

$$u(k\Delta, t) = u_k(t), \tag{15.6.2}$$

and $k(x) = b_k$ to avoid confusing notation, we see that (15.6.1) becomes a set of linear ordinary differential equations

$$\delta^2 b_k \frac{du_k}{dt} = (u_{k+1} + u_{k-1} - 2u_k). \tag{15.6.3}$$

An advantage of this approach lies in the fact that a nonlinear partial differential equation such as (15.5.4) leads directly to a corresponding set of nonlinear ordinary differential equations

$$b_k \frac{du_k}{dt} = \left(\frac{u_{k+1} + u_{k-1} - 2u_k}{\delta^2}\right) + u_k^2, \tag{15.6.4}$$

with the appropriate initial conditions.

15.7. Preservation of Properties

As pointed out above, there are many different types of difference approximations, each with advantages and disadvantages. Can we construct suitable relations which preserve known properties such as nonnegativity, boundedness, and so forth? Furthermore, it would be quite convenient for analytic purposes to have this obviously the case.

Consider, for example, the diffusion equation

$$u_t = u_{xx}, \qquad 0 < x < 1, \quad t > 0, \tag{15.7.1}$$

where

$$\begin{aligned} u(0, t) &= u(1, t) = 0, \qquad & t > 0, \\ u(x, 0) &= g(x), & 0 < x < 1. \end{aligned} \tag{15.7.2}$$

From the property of the associated Green's function we know that if $g \geqslant 0$, then $u \geqslant 0$. The difference approximations of the foregoing sections do not, however, make this immediately clear. If, on the other hand we write

$$v\left(x, t + \frac{\Delta^2}{2}\right) = \frac{v(x - \Delta, t) + v(x + \Delta, t)}{2}, \tag{15.7.3}$$

where $t = 0, \Delta^2/2, \ldots,$ and $x = 0, \Delta, 2\Delta, \ldots, N\Delta = 1$, then it is obvious that $0 \leqslant v(x, 0) \leqslant m$ entails the same relation for $v(x, t)$ for $t > 0$.

The relation in (15.7.3), however, possesses the disadvantages previously discussed, namely relatively poor accuracy together with a long computing time. Thus, it is interesting to see if we can obtain more efficient formulas which still preserve the nonnegativity and boundedness. Let us write

$$v(x, t + \Delta) = \sum_{i=1}^{n} a_i(v(x + \delta_i, t) + v(x - \delta_i, t)), \tag{15.7.4}$$

where the a_i and δ_i are positive values to be determined, and ask that this relation be accurate to order Δ^r, for some integer $r \geqslant 1$, for v satisfying the relation $v_t = v_{xx}$. The values of $v(x + \delta_i, t)$ are obtained via interpolation or the use of a spline or polynomial approximation. Expanding, we have

$$\begin{aligned} v(x, t + \Delta) &= v(x, t) + \Delta v_t + \frac{\Delta^2}{2} v_{tt} + \cdots, \\ v(x + \delta_i, t) + v(x - \delta_i, t) &= 2v(x, t) + \delta_i^2 v_{xx} + \frac{2\delta_i^4}{4!} v_{xxxx} + \cdots. \end{aligned} \tag{15.7.5}$$

If $v(x, t)$ is to satisfy (15.7.1), we must have the relations

$$2 \sum_{i=1}^{n} a_i = 1, \qquad \sum_{i=1}^{n} a_i \delta_i^2 = \Delta, \qquad \frac{2}{4!} \sum_{i=1}^{n} a_i \delta_i^4 = \frac{\Delta^2}{2}, \tag{15.7.6}$$

and so on.

The third relation is a consequence of the fact that $v_t = v_{xx}$ implies

$$v_{tt} = v_{xxxx} \; ; \tag{15.7.7}$$

the further relations are derived similarly.

We see then that we face a moment problem. It is necessary to determine what constraints on r, n and the c_k allow us to solve a system of equations of the form

$$\sum_{i=1}^{n} a_i \delta_i^{2k} = c_k , \qquad k = 0, 1, 2,..., r, \tag{15.7.8}$$

with δ_i , $a_i \geqslant 0$, $i = 1, 2,..., n$. Fortunately, problems of this nature have been intensively studied. References will be found at the end of the chapter.

Exercise

1. Determine sets of a_i and δ_i for $r = 2, 3$, and 4.

15.8. Orthonormal Expansions and Infinite Systems

In place of an approximation method based upon finite differences, we can employ an expansion of the solution in an orthonormal series. Consider, for example, the equation

$$u_t = u_{xx} + u^2, \qquad u(x, 0) = g(x), \tag{15.8.1}$$

where $g(x)$ is periodic of period 2π. We may just as well put the π here as someplace else.

Since we are looking for a solution $u(x, t)$ which is periodic of period 2π for $t > 0$, we can write

$$u(x, t) = \sum_n u_n(t) \, e^{inx}, \tag{15.8.2}$$

where $-\infty < n < \infty$.

Substitution in (15.8.1) and equating of coefficients of e^{inx} leads to the relations

$$u_n' = -n^2 u_n + \sum_{k+\ell=n} u_k u_\ell , \qquad u_n(0) = g_n , \quad -\infty < n, k, l < \infty \tag{15.8.3}$$

where the initial conditions are determined by the Fourier expansion $g(x) \sim \sum_n g_n e^{inx}$.

Since this is an infinite system of ordinary differential equations, it is not in a form immediately suitable either for a digital computer, nor for various analytic approximation methods.

In subsequent sections we will discuss the question of existence and uniqueness of solutions of infinite linear systems of differential equations under various assumptions. The methods described there can be extended to treat various types of infinite nonlinear systems. In what immediately follows we pursue a different route. We suppose that we have established the existence and uniqueness of solution of the infinite system in an appropriate class of functions and are concerned solely with a feasible method for obtaining the solution.

15.9. Truncation

In order to obtain an algorithm which can be used for computational purposes, we truncate. By this we mean that we set

$$u_n = 0, \qquad |n| > N. \tag{15.9.1}$$

This is, of course, only one method of replacing infinite by finite processes. We shall discuss some others below.

Under this assumption (15.8.3) yields the finite system

$$w_n' = -n^2 w_n + \sum_{k+\ell=n} w_k w_\ell, \qquad w_n(0) = g_n, \quad |n| \leqslant N. \tag{15.9.2}$$

In the sum on the right we allow only values of k and l for which $|k|$, $|l| \leqslant N$.

The numerical solution of systems of the foregoing type can readily be carried out for reasonably large values of N at the present time. Furthermore, since this capability is expanding rapidly, the question of justifying this simple, direct approach is of great importance.

Let us introduce the function

$$u^{(N)}(x, t) = \sum_{|n| \leqslant N} w_n(t)\, e^{inx}. \tag{15.9.3}$$

We wish to show that $u^{(N)} \to u$ as $N \to \infty$ where u is the solution of (15.8.1), assumed for the moment to exist. This will be the case if $\| g \| = \max_x | g(x)|$ is sufficiently small, as we have indicated in the exercises in Chapter 10.

15.10. Associated Equation

Let us introduce some simplifying notation. Let f be a function possessing the Fourier expansion

$$f \sim \sum_n f_n e^{inx}, \tag{15.10.1}$$

and let the linear operation $s_N(f)$ be defined by the relation

$$s_N(f) = \sum_{|n| \leqslant N} f_n e^{inx}. \tag{15.10.2}$$

This is to say that $s_N(f)$ is a truncation operator which replaces the function f by the Nth partial sum of its Fourier series.

Returning to the function $u^{(N)}$ defined in (15.9.3), we see that (15.9.2) is equivalent to the relation

$$\sum_{|n| \leqslant N} u_n'(t)\, e^{inx} = - \sum_{|n| \leqslant N} n^2 u_n(t)\, e^{inx} + \sum_{|n| \leqslant N} \Big(\sum_{k+\ell=n} u_k(t)\, u_\ell(t) \Big)\, e^{inx} \tag{15.10.3}$$

(where again $|k|$, $|l| \leqslant N$), which may be written more simply as

$$\frac{\partial u^{(N)}}{\partial t} = \frac{\partial^2 u^N}{\partial x^2} + s_N((u^{(N)})^2). \tag{15.10.4}$$

The initial condition is

$$u^{(N)}(x, 0) = s_N(g). \tag{15.10.5}$$

Let

$$r_N(f) = f - s_N(f). \tag{15.10.6}$$

Then (15.10.4) becomes

$$\frac{\partial u^{(N)}}{\partial t} = \frac{\partial^2 u^{(N)}}{\partial x^2} + (u^{(N)})^2 - r_N[(u^{(N)})^2]. \tag{15.10.7}$$

We see, therefore, that the effect of the truncation is to replace the original equation in (15.8.1) by the equation

$$\frac{\partial v}{\partial t} = \frac{\partial^2 v}{\partial x^2} + v^2 - r_N(v^2), \tag{15.10.8}$$

subject to the initial condition $v(x, 0) = g - r_N(g)$.

Since we expect $r_N(g)$ and $r_N(v^2)$ to approach zero as $N \to \infty$, it is plausible that $\lim_{N\to\infty} u^{(N)} = u$. In any case, we have achieved our first objective of showing that the problem of validating the truncation procedure is equivalent to discussing the behavior of a perturbed form of the original equation.

15.11. Discussion of Convergence of $u^{(N)}$

The problem as posed above cannot be treated in as simple and direct a fashion as desired because of the idiosyncrasies of the operator s_N. As we know from the theory of Fourier series, we do not have a relation of the form

$$\| s_N(f)\| \leqslant k \| f \|, \tag{15.11.1}$$

for some constant k independent of N. Here we are employing the norm,

$$\| f \| = \max_x | f |. \tag{15.11.2}$$

Nor can we assert without imposing more strenuous conditions than continuity that

$$\| r_N(f)\| \leqslant \epsilon \| f \| \tag{15.11.3}$$

for $N \geqslant N(\epsilon)$.

If we do impose conditions on g sufficient to establish these estimates, there is some effort required in showing that these same conditions are maintained when we pursue a method of successive approximations required to treat the existence and uniqueness of the solution of (15.10.8).

Let us then introduce a slight modification of the original truncation scheme, inessential as far as application is concerned, which effectively bypasses these problems. It introduces some interesting problems in connection with the treatment of other classes of partial differential equations which, however, we will not pursue here.

15.12. The Fejer Sum

Let us recall the classical result of Fejer concerning the summability of Fourier series. If we introduce the operation

$$F_N(f) = \frac{s_0(f) + \cdots + s_N(f)}{N+1} = \sum_{|k| \leqslant N} \left(1 - \frac{| k |}{(N+1)}\right) f_k e^{ikx}, \tag{15.12.1}$$

the $(C, 1)$-sum of this series, we know that $F_N(f)$ converges uniformly to f as $N \to \infty$, provided that $f(x)$ is continuous in $[0, 2\pi]$, and, furthermore, that

$$\|F_N(f)\| \leqslant \|f\|, \qquad (15.12.2)$$

with the norm defined as in (15.11.2).

15.13. Modified Partial Differential Equation

Let us then replace (15.10.8) by the equation

$$\frac{\partial w}{\partial t} = \frac{\partial^2 w}{\partial x^2} + w^2 - \rho_N(w^2) \qquad (15.13.1)$$

where

$$\rho_N(f) = f - F_N(f), \qquad (15.13.2)$$

and use the initial condition

$$w(x, 0) = F_N(g). \qquad (15.13.3)$$

It is now a straightforward matter to carry through a proof that

$$\| u - w \| \leqslant \epsilon \qquad (15.13.4)$$

provided that $\| g \|$ is sufficiently small, a condition we require in any case for the existence of a solution of (15.8.1) for all $t \geqslant 0$, and provided that $N \geqslant N(\epsilon)$.

15.14. Modified Truncation

Let us now examine the form of the solution (15.13.1). Write

$$w^{(N)} = \sum_{|n| \leqslant N} w_n(t)\, e^{inx} \qquad (15.14.1)$$

and substitute into (15.12.1). We obtain the finite system of equations

$$w_n'(t) = -n^2 w_n + \left(1 - \frac{|n|}{N+1}\right) \sum_{k+\ell=n} w_k w_\ell\,, \qquad |k|, |l| \leqslant N, \quad (15.14.2)$$

for $n = 0, \pm 1, \ldots, \pm N$, with the initial conditions

$$w_n(0) = \left(1 - \frac{|n|}{N+1}\right) g_n\,, \qquad n = 0, \pm 1, \ldots, \pm N. \qquad (15.14.3)$$

The convergence of $w^{(N)}$ to u as $N \to \infty$ may not be as fast as that of $u^{(N)}$. However, other desirable properties such as boundedness and nonnegativity are guaranteed.

Exercises

1. Consider the equation $u_t = u_{xx} + k(x)u$, where $u(x, 0) = g(x)$ and $g(x)$ and $k(x)$ are periodic of period 2π. Show that $g(x) \geqslant 0$ ensures that $u \geqslant 0$.

2. Write

$$u(x, t) = \sum_n u_n(t)\, e^{inx}, \qquad k(x) = \sum_n k_n e^{inx}, \qquad g(x) = \sum_n g_n e^{inx}.$$

Show that this leads to the infinite system

$$u_n' = -n^2 u_n + \sum_{r+s=n} k_r u_s\,, \qquad u_n(0) = g_n\,.$$

3. Truncation yields the finite system

$$v_n' = -n^2 v_n + \sum_{r+s=n} k_r v_s\,, \qquad v_n(0) = g_n\,, \quad n = 0, \pm 1, ..., \pm N,$$

with $v_n = 0$ for $|\, n\,| > N$. Can we assert that

$$v^{(N)}(x, t) = \textstyle\sum_{|n| \leqslant N} v_n e^{inx}$$

is nonnegative if $g(x) \geqslant 0$?

4. What about the solution of $w_t = w_{xx} + F_N(k(x)w)$, $w(x, 0) = F_N(g)$? Consider first, the case where $k(x) \geqslant 0$.

5. Consider the general case by means of the equation $z_t = z_{xx} + F_N((a + k(x))z)$, $z(x, 0) = F_N(g)$, where a is chosen so that $a + k(x) > 0$.

15.15. Infinite Systems of Ordinary Differential Equations

Let us now examine the existence and uniqueness of solutions of infinite systems of linear differential equations of the form

$$\frac{dx_i}{dt} = \sum_{j=1}^{\infty} a_{ij} x_j\,, \qquad x_i(0) = c_i\,, \tag{15.15.1}$$

$i = 1, 2, \ldots$. Using the appropriate vector and matrix, we may write this in the form

$$\frac{dx}{dt} = Ax, \qquad x(0) = c. \tag{15.15.2}$$

We begin by introducing some norms: write

$$N_p(x) = \left(\sum_{i=1}^{\infty} | x_i |^p\right)^{1/p},$$
$$N_q(A) = \left[\sum_{i=1}^{\infty} \left(\sum_{j=1}^{\infty} | a_{ij} |^q\right)^{p-1}\right]^{1/p}, \tag{15.15.3}$$

where we take $p > 1$ and

$$\frac{1}{p} + \frac{1}{q} = 1. \tag{15.15.4}$$

Here x is the infinite dimensional vector whose components are the x_i and A is the infinite-dimensional matrix whose elements are the a_{ij}, which may be functions of t.

Let us note that Hölder's inequality shows that

$$N_p(Ax) \leqslant N_q(A)\, N_p(x), \tag{15.15.5}$$

and Minkowski's inequality that

$$N_p(x + y) \leqslant N_p(x) + N_p(y). \tag{15.15.6}$$

The case $p = 2$ is particularly interesting. Here

$$N_2(A) = \left(\sum_{i,j=1}^{\infty} | a_{ij} |^2\right)^{1/2}. \tag{15.15.7}$$

If $p = 1$, we set

$$N_1(x) = \sum_{i=1}^{\infty} | x_i |, \qquad N_1(A) = \sum_{i,j=1}^{\infty} | a_{ij} |. \tag{15.15.8}$$

We wish to demonstrate:

Consider the differential equation

$$\frac{dx}{dt} = Ax, \qquad x(0) = x_0 . \tag{15.15.9}$$

If

$$N_p(x_0) < \infty, \qquad \int_0^t N_q(A)\, dt_1 < \infty \tag{15.15.10}$$

for $t \leqslant t_0$, *there is a unique solution of* (15.15.9) *for which* $N_p(x) < \infty$ *for* $t \leqslant t_0$.

15.16. Proof of Theorem

We begin with a proof of existence and then provide a proof of uniqueness. As usual, we employ the method of successive approximations. Write

$$x = x_0 + \int_0^t Ax\, dt_1 \tag{15.16.1}$$

and consider the sequence of functions defined by

$$x_0 = x_0\,, \qquad x_{n+1} = x_0 + \int_0^t Ax_n\, dt_1\,. \tag{15.16.2}$$

The first task is to show that each x_{n+1} is such that $N_p(x_{n+1}) < \infty$. For this purpose, we use the inequalities

$$\begin{aligned} N_p(x_{n+1}) &\leqslant N_p(x_0) + N_p\left(\int_0^t Ax_n\, dt_1\right) \\ &\leqslant N_p(x_0) + \int_0^t N_p(Ax_n)\, dt_1 \\ &\leqslant N_p(x_0) + \int_0^t N_q(A)\, N_p(x_n)\, dt_1\,. \end{aligned} \tag{15.16.3}$$

(Set $q = 1$, if $p = 1$).

Let us now establish inductively that

$$N_p(x_n) \leqslant N_p(x_0) \exp\left(\int_0^t N_q(A)\, dt_1\right). \tag{15.16.4}$$

This obviously holds for $n = 0$. Combining the result for n with (15.16.3) we get the result for $n + 1$.

It remains to establish the convergence of the sequence $\{x_n\}$. We subtract

$$x_{n+1} - x_n = \int_0^t A(x_n - x_{n-1})\, dt_1\,, \qquad x_1 - x_0 = \int_0^t Ax_0\, dt_1\,. \tag{15.16.5}$$

Then it follows that

$$N_p(x_1 - x_0) \leqslant N_p(x_0) \int_0^t N_p(A)\, dt_1 \tag{15.16.6}$$

and thus by iteration,

$$\begin{aligned} N_p(x_{n+1} - x_n) &\leqslant \int_0^t N_q(A)\, N_p(x_n - x_{n-1})\, dt_1 \\ &\leqslant \int_0^t N_q(A)\, N_p(x_0) \frac{[\int_0^t N_q(A)\, dt_2]^n}{n!}\, dt_1 \\ &\leqslant N_p(x_0) \frac{[\int_0^t N_q(A)\, dt_1]^{n+1}}{(n+1)!}\,. \end{aligned} \tag{15.16.7}$$

Thus the series $\sum_{n=0}^{\infty} N_p(x_{n+1} - x_n)$ converges uniformly in any finite t-interval. Since $N_p(x_n - x_m) \leqslant \sum_{k=m+1}^{n} N_p(x_k - x_{k-1})$, the Cauchy condition will be satisfied, and there exists a vector x with $N_\phi(x) < \infty$, such that $N_p(x - x_n) \to 0$ as $n \to \infty$, uniformly in any finite t-interval. Using (15.16.4) we see that

$$N_p(x) \leqslant N_p(x_0) \exp\left(\int_0^t N_q(A)\, dt_1\right). \tag{15.16.8}$$

Since

$$x_{n+1} = x_0 + \int_0^t A x_n\, dt_1 \tag{15.16.9}$$

and x_n converges uniformly to x, we have

$$x = x_0 + \int_0^t A x\, dt_1\,. \tag{15.16.10}$$

Thus x satisfies the differential equation. To show the uniqueness of solution, we let y be another solution of the differential equation such that $N_p(y) < \infty$. Then

$$x - y = \int_0^t A(x - y)\, dt_1\,,$$

$$N_p(x - y) \leqslant N_p\left(\int_0^t A(x - y)\, dt_1\right) \leqslant \int_0^t N_q(A)\, N_p(x - y)\, dt_1\,, \tag{15.16.11}$$

whence

$$N_p(x - y) \leqslant \exp\left(\int_0^t N_q(A)\, dt_1\right). \tag{15.16.12}$$

Substituting this relation in (15.16.11) we obtain

$$N_p(x-y) \leqslant \int_0^t N_q(A)\left[\exp\left(\int_0^t N_q(A)\,dt_1\right)\right] dt$$

$$\leqslant \left[\exp \int_0^t N_q(A)\,dt_1\right]\left[\int_0^t N_q(A)\,dt_1\right]^2 \Big/ 2! \qquad (15.16.13)$$

Substituting again, one obtains

$$N_p(x-y) \leqslant \left[\exp\left(\int_0^t N_q(A)\,dt_1\right)\right]\left[\int_0^t N_q(A)\,dt_1\right]^2 \Big/ 2! \qquad (15.16.14)$$

Continuing in this way, we see that $N_p(x-y) \equiv 0$, and that $x \equiv y$.

Exercises

1. Apply the foregoing to study the equation

$$u_t = u_{xx} + k(x)u, \qquad u(x, 0) = g(x), \quad u(0, t) = u(1, t) = 0, \quad t > 0.$$

2. Consider next the equation

$$u_t = u_{xx} + u^2, \qquad u(x, 0) = g(x), \quad u(0, t) = u(1, t) = 0, \quad t > 0.$$

15.17. The "Principe des Réduites"

Let us next consider the equation

$$\frac{dx}{dt} = Ax, \qquad x(0) = x_0, \qquad (15.17.1)$$

and an associated approximate equation

$$\frac{dx^{(n)}}{dt} = A_n x^{(n)}, \qquad x^{(n)}(0) = x_0^{(n)}, \qquad (15.17.2)$$

where A_n is the nth section of the infinite matrix, defined by the relations

$$A_n = (a_{ij}^{(n)}), \qquad (15.17.3)$$

where

$$a_{ij}^{(n)} = a_{ij} \qquad (0 \leqslant i, j \leqslant n),$$

$$a_{ij}^{(n)} = 0 \qquad (i, j > n),$$

and x_n is the vector composed of the first n components of x, with the remaining components zero; $x_0^{(n)}$ defined similarly.

A question of importance, particularly for computational studies, is whether $x^{(n)}$ converges as $n \to \infty$ and, if so, whether it converges to a solution of (15.17.1). If $x^{(n)} \to x$ as $n \to \infty$, the equation may be said to satisfy the "principe des reduites," a term used by F. Riesz to denote the corresponding procedure for infinite systems of linear algebraic equations. We shall prove the following:

Under the conditions of the theorem of Sec. 15 *concerning A and x_0, the "principe des reduites" is valid.*

PROOF. Converting each equation of the form (15.17.2) into an integral equation of the type previously used, we have

$$\begin{aligned} x^{(n)} - x^{(m)} &= x_0^{(n)} - x_0^{(m)} + \int_0^t [A_n x^{(n)} - A_m x^{(m)}]\, dt_1 \\ &= x_0^{(n)} - x_0^{(m)} + \int_0^t A_n(x^{(n)} - x^{(m)})\, dt_1 + \int_0^t (A_n - A_m)\, x^{(m)}\, dt_1 . \end{aligned} \tag{15.17.4}$$

Thus

$$\begin{aligned} N_p(x^{(n)} - x^{(m)}) \leqslant N_p(x_0^{(n)} - x_0^{(m)}) &+ \int_0^t N_q(A_n)\, N_p(x^{(n)} - x^{(m)})\, dt_1 \\ &+ \int_0^t N_p(x^{(m)})\, N_q(A_n - A_m)\, dt_1 . \end{aligned} \tag{15.17.5}$$

Since, as we have seen above,

$$N_p(x^{(m)}) \leqslant N_p(x_0^{(m)}) \exp\left(\int_0^t N_q(A_n)\, dt_1\right) \leqslant N_p(x_0) \left(\exp \int_0^t N_q(A_n)\, dt_1\right), \tag{15.17.6}$$

and $N_q(A_n) \leqslant N_q(A)$, we have

$$\begin{aligned} N_p(x^{(n)} - x^{(m)}) \leqslant N_p(x_0^{(n)} - x_0^{(m)}) &+ \int_0^t N_q(A)\, N_p(x^{(n)} - x^{(m)})\, dt_1 \\ &+ N_p(x_0) \left(\exp \int_0^t N_q(A)\, dt_1\right) \int_0^t N_q(A_n - A_m)\, dt_1 . \end{aligned} \tag{15.17.7}$$

Let us now restrict t to the fixed interval $(0, t_2)$, $t_2 > 0$.

Then for $0 \leqslant t \leqslant t_2$,

$$N_p(x^{(n)} - x^{(m)}) \leqslant N_p(x_0^{(n)} - x_0^{(m)}) + N_p(x_0) \cdot \left(\exp \int_0^{t_2} N_q(A)\, dt_1\right) \int_0^{t_2} N_q(A_n - A_m)\, dt_1 + \int_0^t N_q(A)\, N_p(x^{(n)} - x^{(m)})\, dt_1 \tag{15.17.8}$$

This inequality has the form

$$f(t) \leqslant c + \int_0^t f(t_1)\, g(t_1)\, dt_1 \qquad (|\, c\,| \neq 0), \tag{15.17.9}$$

where $f(t) = N_p(x^{(n)} - x^{(m)})$. We now use the fundamental lemma to obtain the following relation:

$$N_p(x^{(n)} - x^{(m)}) \leqslant \left[N_p(x_0^{(n)} - x_0^{(m)}) + N_p(x_0) \cdot \left(\exp \int_0^{t_2} N_q(A)\, dt_1\right) \int_0^{t_2} N_q(A_n - A_m)\, dt_1\right] \cdot \left[\exp \int_0^t N_q(A)\, dt_1\right]. \tag{15.17.10}$$

For n and m sufficiently large, $N_p(x_0^{(n)} - x_0^{(m)}) \leqslant \epsilon$, and since $N_q(A_n - A_m) \to 0$ as $n, m \to \infty$ and $N_q(A_n - A_m) \leqslant N_q(A)$, we have, for $n, m \geqslant n_0(\epsilon)$,

$$N_p(x^{(n)} - x^{(m)}) \leqslant c_1 \epsilon \exp\left(\int_0^{t_2} N_q(A)\, dt_1\right) \leqslant c_2 \epsilon. \tag{15.17.11}$$

Thus a vector x exists with $N_p(x) < \infty$, such that $N_p(x^{(n)} - x) \to 0$ as $n \to \infty$. Since

$$x^{(n)} = x_0^{(n)} + \int_0^t A_n x^{(n)}\, dt_1, \tag{15.17.12}$$

and $N_q(A - A_n) \to 0$ as $n \to \infty$, we have

$$x = x_0 + \int_0^t Ax\, dt_1, \tag{15.17.13}$$

and thus x is a solution of (15.17.1), and, as we know, the unique solution.

Exercise

1. Show that the restriction that $N_p(x) < \infty$ is essential. (Consider, for example, the system

$$y_k' = \frac{y_{k+1}}{(k+1)^2}, \qquad y_k(0) = 0,$$

$k = 1, 2, \ldots$. A nontrivial solution is obtained by taking $y_1(t) = \exp(-1/t^2)$, with $y_k(t)$ for $k \geqslant 2$ determined by the above. Then $N_p(y)$ is finite for no value of p for any $t > 0$.)

15.18. Random Walk

A prolific source of infinite systems of differential equations is the theory of stochastic processes. Consider for example, a system S which may be in any of a denumerable set of states at any particular time t. Without loss of generality, label these states 1, 2,... . Let

$$p_i(t) = \text{the probability that } S \text{ is in state } i \text{ at time } t; \quad i = 1, 2, \ldots, \quad t \geqslant 0. \tag{15.18.1}$$

Furthermore, let Δ be an infinitesimal and to terms which are $o(\Delta)$, let

$$\begin{aligned} a_{ij}\Delta &= \text{the probability that the system jumps from state } j \text{ to state } i \text{ in the } t\text{-interval } [t, t+\Delta], \quad j \neq i, \\ 1 - a_{ii}\Delta &= \text{the probability that the system remains in state } i \text{ over } [t, t+\Delta]. \end{aligned} \tag{15.18.2}$$

Then the usual reasoning yields the relations

$$p_i(t+\Delta) = (1 - a_{ii}\Delta)\, p_i(t) + \Delta \sum_{j \neq i} a_{ij} p_j(t), \qquad i = 1, 2, \ldots, \tag{15.18.3}$$

where the right-hand side is accurate to terms in $o(\Delta)$. Passing to the limit in the customary fashion, we obtain an infinite system of linear differential equations

$$p_i'(t) = -a_{ii} p_i(t) + \sum_{j \neq i} a_{ij} p_j(t), \qquad p_i(0) = c_i, \tag{15.18.4}$$

$i = 1, 2, \ldots$. Here $a_{ij} \geqslant 0$, $c_i \geqslant 0$, and

$$\sum_{i=1}^{\infty} c_i = 1, \qquad \sum_{i \neq j} a_{ij} = a_{jj}, \qquad j = 1, 2, \ldots . \tag{15.18.5}$$

We now take these equations as the defining equations for a continuous stochastic process.

The question of existence and uniqueness of solution of systems of this type is not without subtleties. Here we wish to concentrate only on the validity of truncation of different types, and even then solely on certain simple aspects.

15.19. The "Principe des Reduites"

Consider the N functions, $q_1, q_2, \ldots, q_N$, obtained as the solution of the finite system

$$q_i'(t) = -a_{ii}q_i + \sum_{1 \leqslant j \leqslant N}' a_{ij}q_j(t), \qquad q_i(0) = c_i, \tag{15.19.1}$$

$i = 1, 2, \ldots, N$, where the prime on the right-hand side indicates that the value $j = i$ is omitted. Using the result of Sec. 3.10, we can assert that the solution is nonnegative, i.e., that $q_i(t) \geqslant 0$ for $t \geqslant 0$. However, the $q_i(t)$ are in general not the probabilities for a system with N allowable states, $1, 2, \ldots, N$, since we do not possess the required normalization $\sum_{i=1}^{n} q_i = 1$. As a matter of fact, we have

$$\frac{d}{dt}\left(\sum_{i=1}^{N} q_i(t)\right) = \sum_{j=1}^{N}\left(\sum_{1 \leqslant j \leqslant N}' a_{ij} - a_{jj}\right) q_j \leqslant 0, \tag{15.19.2}$$

since $q_j \geqslant 0$ and we have, by virtue of (15.18.5),

$$\sum_{1 \leqslant j \leqslant N}' a_{ij} - a_{jj} \leqslant 0. \tag{15.19.3}$$

If any of the a_{ij} are positive for $1 \leqslant i \leqslant N$, $j > N$, then (15.19.2) becomes

$$\frac{d}{dt}\left(\sum_{i=1} q_i(t)\right) < 0. \tag{15.19.4}$$

Hence $\sum_{i=1} q_i(t)$ is strictly decreasing as t increases. This does show, however, that each $q_i(t)$ is uniformly bounded as $t \to \infty$.

$$0 \leqslant q_i(t) \leqslant 1. \tag{15.19.5}$$

15.20. Monotone Convergence

Let us now write $q^{(N)}(t)$, $i = 1, 2, \ldots, N$, to indicate explicitly the dependence on N. Then we can assert that

$$q_i^{(N)} \leqslant q_i^{(N+1)}. \tag{15.20.1}$$

The proof follows once again from the results in Sec. 3.10, Volume I. We have

$$\frac{d}{dt}(q_i^{(N)}) = -a_{ii}q_i^{(N)} + \sum_{1\leqslant j\leqslant N}' a_{ij}q_j^{(N)}, \qquad q_i^{(N)}(0) = c_i, \quad (15.20.2)$$

$i = 1, 2,..., N$, and

$$\frac{d}{dt}(q_i^{(N+1)}) = -a_{ii}q_i^{(N+1)} + \sum_{1\leqslant j\leqslant N+1}' a_{ij}q_j^{(N+1)}, \qquad q_i^{(N+1)}(0) = c_i, \quad (15.20.3)$$

$i = 1, 2,..., N + 1$. Hence

$$\frac{d}{dt}(q_i^{(N+1)} - q_i^{(N)}) = -a_{ii}(q_i^{(N+1)} - q_i^{(N)}) + \sum_{1\leqslant j\leqslant N} a_{ij}(q_j^{(N+1)} - q_j^{(N)}) + a_{i,N+1}q_{N+1}^{(N+1)}, \quad (15.20.4)$$

with $q_i^{(N+1)}(0) - q_i^{(N)}(0) = 0$. Hence, from the stated result

$$q_i^{(N+1)} - q_i^{(N)} \geqslant 0. \quad (15.20.5)$$

Since the $q_i^{(N)}$ are uniformly bounded, we see that

$$\lim_{N\to\infty} q_i^{(N)} = q_i, \qquad i = 1, 2,..., \quad (15.20.6)$$

where we have constantly suppressed the t-dependence in both q_i and $q_i^{(N)}$.

Use of the same comparison technique shows that

$$q_i \leqslant p_i, \qquad i = 1, 2,..., \quad (15.20.7)$$

whenever the system in (15.19.4) possesses a nonnegative solution.

Exercises

1. Do the $q_i(t)$ satisfy (15.20.4)?

15.21. Existence of a Solution

Let us assume that $\sum_j a_{ij}$ converges for each i, and that (15.18.4) possesses a unique solution $p_i(t)$ representing a set of probabilities for $t \geqslant 0$. Write

$$p_i' = -a_{ii}p_i + \sum_{1\leqslant j\leqslant N}' a_{ij}p_j + \sum_{j>N} a_{ij}p_j, \qquad p_i(0) = c_i, \quad (15.21.1)$$

$i = 1, 2,\dots, N$. Considering $\sum_{j>N} a_{ij}p_j$ as a forcing term, we can write

$$p_i = q_i^{(N)} + \sum_{j=1}^{N} \int_0^t k_{ij}(t - t_1) \left[\sum_{k>N} a_{jk}p_k(t_1) \right] dt_1 . \qquad (15.21.2)$$

Here $p(t) = (k_{ij}(t))$ is the solution of the vector equation

$$\frac{dP}{dt} = BP, \qquad P(0) = I, \qquad (15.21.3)$$

where

$$B = \begin{bmatrix} -a_{11} & a_{12} & \cdots & a_{1N} \\ a_{21} & -a_{22} & \cdots & a_{2N} \\ \vdots & \vdots & & \vdots \\ a_{N1} & a_{N2} & & -a_{NN} \end{bmatrix}. \qquad (15.21.4)$$

We see that $1 \geqslant k_{ij}(t) \geqslant 0$ for $t \geqslant 0$, $1 \leqslant i$, $j \leqslant N$. Hence, since $\sum_k p_k = 1$, we have

$$| p_i - q_i^{(N)} | \leqslant \sum_{j=1}^{N} \int_0^t \left(\sum_{k>N} a_{jk} \right) dt_1 \leqslant t \sum_{j=1}^{N} \sum_{k>N} a_{jk} . \qquad (15.21.5)$$

Since $\sum_j a_{ij}$ converges, by hypothesis, we see that

$$\lim_{N\to\infty} q_i^{(N)} = p_i , \qquad (15.21.6)$$

not necessarily uniformly in either i or t.

A particularly important case of the above condition is that where $a_{ij} = 0$ for $j \geqslant j(i)$.

Exercise

1. Under what additional conditions will the convergence be uniform in t and i?

15.22. Closure of the Process

In place of writing down the exact equations for the original process and then approximating to the equations as we did above, we can begin by approximating to the process and then using the exact equations for the approximating process. Choosing convenient approximate processes guided by the underlying physical process we can thus obtain finite systems of equations.

An important point to stress is that an approximate process always yields equations which have certain desired physical attributes. On the other hand, approximate equations may not correspond to any physical process.

In order to illustrate this approach, let us replace the original random walk process described above by one in which there are only N states. Consider then the finite system of equations

$$r_i' = -b_{ii}r_i + \sum_{j=1}^{N} b_{ij}r_j, \qquad r_i(0) = c_i, \tag{15.22.1}$$

where we use the modified probabilities

$$b_{ij} = a_{ij}, \qquad i = 1, 2,..., N-1,$$

$$b_{Nj} = a_{jj} - \sum_{i=1}^{N-1} a_{ij}. \tag{15.22.2}$$

In this case, we modify only the equation for r_N. It follows immediately that the $r_i(t)$ are now probabilities for all $t \geqslant 0$. Since

$$a_{jj} = \sum_{i\neq j}{}' a_{ij} = \sum_{i=1}^{N} a_{ij} + \sum_{i>N} a_{ij}, \tag{15.22.3}$$

we see that

$$a_{Nj} = a_{jj} - \sum_{i=1}^{N-1} a_{ij} - \sum_{i>N} a_{ij} = b_{Nj} - \sum_{i>N} a_{ij}. \tag{15.22.4}$$

Hence,

$$| a_{Nj} - b_{Nj} | \leqslant e_j^{(N)}, \tag{15.22.5}$$

where $e_j^{(N)} \to 0$ as $N \to \infty$ for each j.

Exercise

1. Establish conditions for convergence of the $r_j^{(N)}$ to the p_i.

Miscellaneous Exercises

1. Let the class C of vector functions $x(t) = (x_1(t), x_2(t),...)$ over the interval $[0, T]$ be defined as follows:

   (1) $x_i(t)$ is absolutely continuous,

(2) $\| x(t)\| = \sum_{i=1}^{\infty} | x_i(t)| \leqslant c < \infty$,

(3) $\| x(t)\|$ is continuous for $0 \leqslant t \leqslant T$.

Then $\dot{x} = Ax$, $x(0) = c$, possesses a unique solution in C for t in $[0, T]$ provided that each $a_{ij}(t)$ is integrable and

(1) $\| c \| < \infty$,

(2) $\sum_{j=1}^{\infty} | a_{ij}(t)| \leqslant a < \infty$ for all i,

(3) $\sum_{i=1} | a_{ij}(t)| \leqslant b < \infty$ for all j

for t in $[0, T]$.

2. Establish the "principe des reduites" under the foregoing conditions with uniform convergence. See

   L. Shaw, "Existence and Approximation of Solutions to an Infinite Set of Linear Time Invariant Differential Equations," *SIAM J. Appl. Math.* (to appear).

   L. Shaw, "Solutions for Infinite-Matrix Differential Equations," *J. Math. Anal. Appl.* (to appear).

3. Consider the equation $u_t = uu_x$, $u(x, 0) = g(x)$, $-\infty < x < \infty$. Show that the relation $u(x, t + \Delta) = u(x + u(x, t)\Delta, t)$ is accurate to terms which are $O(\Delta^2)$.

   R. Bellman, I. Cherry, and G. M. Wing, "A Note on the Numerical Integration of a Class of Nonlinear Hyperbolic Equations," *Quart. Appl. Math.*, Vol. 16, 1958, pp. 181–183.

4. Show that $u(x, t + \Delta) = u(x + u(x + u(x, t)\Delta, t)\Delta, t)$ is accurate to $O(\Delta)^3$.

5. Near the singularity of the solution which approximation should yield better numerical results? For a discussion and some numerical results, see

   S. Azen, *Higher Order Approximations to the Computational Solution of Partial Differential Equations*, The RAND Corporation, RM-3917-ARPA, 1964.

   See also

   R. Bellman and K. L. Cooke, "Existence and Uniqueness Theorems in Invariant Imbedding—II, Convergence of a New Difference Algorithm," *J. Math. Anal. Appl.*, Vol. 12, 1965, pp. 247–253.

For some related results, see

R. Bellman, R. Kalaba, and G. M. Wing, "Invariant Imbedding and Neutron Transport Theory—V: Diffusion as a Limiting Case," *J. Math. Mech.*, Vol. 9, 1960, pp. 933–944.

R. Bellman and K. L. Cooke, "On the Limit of Solutions of Differential-Difference Equations as the Retardation Approaches Zero," *Proc. Nat. Acad. Sci.*, Vol. 45, 1959, pp. 1026–1028.)

6. Consider Burgers' equation $u_t + uu_x = \epsilon u_{xx}$ (which may be considered to be a Riccati form of the linear heat equation $v_t = v_{xx}$). Suppose that the initial function $u(x, 0) = g(x)$ is period of period 2π and write

$$u(x, t) = \sum_{-\infty<k<\infty} u_k(t)\, e^{ikx}$$

where $u_k = v_k + iw_k$, v_k and w_k real. Obtain the equations

$$u_n'(t) + \sum_{-\infty<k<\infty} iku_k(t)\, u_{n-k}(t) = -n^2\epsilon u_n(t), \qquad -\infty < n < \infty.$$

7. Consider the truncated version where $|n| \leqslant N$. How do we know what value of N to use in obtaining a numerical solution? See the discussion in

R. Bellman, S. P. Azen, and J. M. Richardson, "On New and Direct Computational Approaches to Some Mathematical Models of Turbulence," *Quart. Appl. Math.*, Vol. 23, 1965, pp. 55–67.

8. Consider the difference approximation

$$\begin{aligned} u(x, t + \Delta) \\ &= \lambda u(x - au(x, t)\,\Delta, t) + \frac{(1-\lambda)}{2}\,[u(x + b\Delta^{1/2}, t) + u(x - b\Delta^{1/2}, t)], \end{aligned}$$

where λ, a, b, Δ are parameters to be determined. How? (See the discussion in the paper cited above.)

9. Consider the equation $u' = -u + u^2$, $u(0) = c$, $|c| \ll 1$. Let $u_n(t) = u^n$, $n = 1, 2, \ldots$. Show that $\{u_n(t)\}$ satisfies the infinite linear system

$$u_n'(t) = -nu_n + nu_{n+1}, \qquad u_n(0) = c^n, \qquad n = 1, 2, \ldots$$

(Carleman linearization).

10. Under what assumptions is this solution unique?

11. Consider the truncated system

$$v_n'(t) = -nv_n + nv_{n+1}, \qquad n = 1, 2, \ldots, N - 1,$$
$$v_N'(t) = -Nv_n,$$

with $v_k(0) = c^k$, $k = 1, 2, \ldots, N$. Write $v_n \equiv v_{n,N}$. Under what conditions do we have convergence, $\lim_{N\to\infty} v_{n,N} = u_n$?

12. Consider the analogous procedure and problem for the vector system $x' = Ax + g(x)$, $x(0) = c$, where $g(x)$ has components which are polynomials in the components of x.

13. If we set $w_n = p_n(u)$ in Exercise 9, where p_n is the nth Legendre polynomial, do we obtain analogous results?

14. Is there any advantage in practice to using one set of functions, $\{u^n\}$, rather than the other, $\{p_n(u)\}$?

15. Consider the equation $u'' + \phi(t)u = 0$ where $\phi(t)$ is continuous and periodic of period 2π. Floquet's theorem asserts that every solution is a linear combination of two solutions of the form $e^{rt}g(t)$ where $g(t)$ is periodic of period 2π. Set $g(t) = \sum_{-\infty<n<\infty} g_n e^{int}$ and obtain an infinite system of linear algebraic equation for the g_n.

16. Use the principe des reduites to obtain approximate values for r and the g_n. This is the method of Hill. See

F. Whittaker and G. M. Watson, *Modern Analysis*, Cambridge Univ. Press, London and New York, 1945.

For an interesting extension to linear equations with almost-periodic coefficients, see

J. M. Abel, "Uniform Almost Orthogonality and the Instabilities of an Almost-Periodic Oscillator," *J. Math. Anal. Appl.*, Vol. 36, 1971, pp. 110–122.

J. M. Abel, "On an Almost Periodic Mathieu Equation," *Quart. Appl. Math.*, Vol. 28, 1970, pp. 205–217.)

17. Consider the partial differential equation

$$\frac{\partial f}{\partial t} = \frac{\partial^2 f}{\partial x^2} + \frac{\partial}{\partial x}(x + ax^3)f, \qquad a > 0, \quad t > 0, \quad -\infty < x < \infty$$
$$f(x, 0) = \delta(x - x_0)$$

which arises in the study of Brownian motion subject to a nonlinear restoring force. Set $u_n(t) = \int_{-\infty}^{\infty} x^n f(x, t)\, dx$, $n = 0, 1, 2, \ldots$. Show that for $n \geqslant 2$ we have

$$u_n'(t) = n(n-1)\, u_{n-2}(t) - nu_n(t) - au_{n+2}\,, \qquad u_n(0) = x_0^{\,n}.$$

18. Using the results of the Miscellaneous Exercises on p. 253 of Volume I obtain a closure which preserves the moment properties. For this and a number of other methods for truncation, see

    R. M. Wilcox and R. Bellman, "Truncation and Preservation of Moment Properties for Fokker-Planck Moment Equations," *J. Math. Anal. Appl.*, Vol. 30, 1970, pp. 592–595.

    See also

    S. Steinberg, "Infinite Systems of Ordinary Differential Equations with Unbounded Coefficients and Moment Problems," *J. Math. Anal. Appl.*, Vol. 41, 1973.

19. Consider the infinite system

$$p_n'(t) = -(\lambda n - \mu n^2)\, p_n(t) + (\lambda(n-1) - \mu(n-1)^2)\, p_{n-1}(t)$$

which arises from a birth-and-death process. Let $u_1(t) = \sum_n np_n(t)$, $u_2(t) = \sum_n n^2 p_n(t)$, be the first and second moments. Show that $u_1'(t) = \lambda u_1(t) - \mu u_2(t)$. Under the assumption that $u_2(t) \cong u_1(t)^2$, obtain the approximate Riccati equation $u'(t) = \lambda u(t) - \mu u(t)^2$.

20. Obtain similar results for the differential equation

$$\frac{\partial w}{\partial t} = b\,\frac{\partial^2 w}{\partial x^2} - \frac{\partial}{\partial x}\,[(ax - cx^3)w],$$

also associated with population growth. For these results, connections with Volterra's theory of "la lutte pour la vie" and further results, see

W. Feller, "Die Grundlegen der Volterraschen Theorie des Kampfes ums Dasein in Wahrscheinlichkeits-theoretischer Behandlung," *Acta Biotheoret. V*, 1939, pp. 12–39.

Bibliography and Comment

15.1. The classic work on infinite systems of linear algebraic equations with reference to work of Fourier, Hill, von Koch and Poincaré is

F. Riesz, *Les Systemes d'Équations Lineaires à une Infinité d'Inconnues*, Paris, 1913.

§15.5. See

R. Varga, *Matrix Iterative Analysis*, Prentice-Hall, Englewood Cliffs, New Jersey, 1962.

§15.6. See

G. Birkhoff and R. S. Varga, "Discretization Errors for Wellset Cauchy Problems. I," *J. Math. and Physics*, Vol. 44, 1965, pp. 1–23.

§15.7. See

T. Motzkin and W. Wasow, "On the Approximation of Linear Elliptic Differential Equations by Difference Equations with Positive Coefficients," *J. Mathematical Phys.*, Vol. 31, 1953, pp. 253–259.

§15.8. This method dates back to Lichtenstein. See

S. Minakshisundaram, "Fourier Ansatz and Nonlinear Parabolic Equations," *J. Indian Math. Soc.*, Vol. 7, 1943, pp. 129–143.

R. Siddiqi, "Zur Theorie der Nichtlinearen Partiellen Differentialgleichungen vom Parabolischen Typus," *Math. Z.*, Vol. 35, 1932, pp. 464–484.

R. Bellman, "On the Existence and Boundedness of Solutions of Nonlinear Partial Differential Equations of Parabolic Type," *Trans. Amer. Math. Soc.*, Vol. 64, 1948, pp. 21–44.

§15.9. For analysis of the validity of this, see

R. Bellman, "On the Validity of Truncation for Infinite Systems of Ordinary Differential Equations Associated with Nonlinear Partial Differential Equations," *Math. and Phys. Sci.* (to appear).

For a discussion of various closure techniques, see

L. C. Leavitt and J. M. Richardson, "Linearized Superposition Theory of a Classical One-Component Plasmic," *Phys. Fluids*, Vol. 10, 1967, pp. 406–413.

§15.15. This was first presented in

R. Bellman, "The Boundedness of Solutions of Infinite Systems of Linear Differential Equations," *Duke Math. J.*, Vol. 14, 1947, pp. 695–706.

See also the papers of Shaw referred to in the Miscellaneous Exercises.

For some applications of infinite systems, see

M. N. Oguztoreli, "On an Infinite System of Differential Equations Occurring in the Degradation of Polymers," *Utilitas Math.*, Vol. 1, 1972, pp. 141–155.

N. Arley and V. Borchsenius, "On the Theory of Infinite Systems of Differential Equations and Their Application to the Theory of Stochastic Processes and the Perturbation Theory of Quantum Mechanics," *Acta Math.*, Vol. 76, 1945, pp. 261–322.

N. Arley, *On the Theory of Stochastic Processes and Their Application to the Theory of Cosmic Radiation*, Copenhagen, 1943.

C. L. Dolph and D. C. Lewis, "On the Application of Infinite Systems of Ordinary Differential Equations to Perturbations of Plane Poiseville Flows," *Quart. Appl. Math.*, Vol. 16, 1958, pp. 97–110.

and the Bellman, Azen, Richardson paper referred to in Exercise 7 of the Miscellaneous Exercises.

§15.18. See

D. R. Cox and H. D. Miller, *The Theory of Stochastic Processes*, Methuen, London, 1965,

where many references and applications are given.

Chapter 16 INTEGRAL AND DIFFERENTIAL QUADRATURE

16.1. Introduction

Classically, a quadrature technique is a numerical procedure for the evaluation of a definite integral in terms of a linear combination of values of the integrand at specific points. A typical formula in the one-dimensional case assumes the form

$$\int_0^1 f(t)\, dt \cong \sum_{i=1}^{N} w_i f(t_i). \tag{16.1.1}$$

We can think of this as a closure technique whereby a transcendental operation, integration in this case, is replaced by function evaluation, which may, of course, require further arithmetic closure, together with the arithmetic operations of addition and multiplication. Consequently, quadrature techniques are fundamental tools in the numerical solution of integral and integro–differential equations by means of a digital computer.

In this concluding chapter we shall first give some examples of this methodology in connection with the numerical inversion of the Laplace transform. Then we shall apply the procedure to a fundamental integro–differential equation of radiative transfer. Finally, we shall use a closely related idea, differential quadrature, in connection with some questions in the identification of systems and the numerical solution of partial differential equations.

16.2. Laplace Transform

In Volume I, Section 1.5, we introduced the Laplace transform

$$L(u) = \int_0^\infty u(t)\, e^{-st}\, dt. \tag{16.2.1}$$

The integral is assumed to converge for $s > 0$. Let us briefly review here some of the features of this operation which make it one of the most powerful of the tools of analysis.

First of all, it is a linear operation,

$$L(c_1u_1 + c_2u_2) = c_1L(u_1) + c_2L(u_2) \tag{16.2.2}$$

for any constants c_1 and c_2 and functions u_1 and u_2 for which $L(u_1)$ and $L(u_2)$ are defined. Secondly, translations are handled with facility,

$$L(u(t + t_1)) = \int_0^\infty u(t + t_1)\, e^{-st}\, dt = \left(\int_{t_1}^\infty u(t)\, e^{-st}\, dt\right) e^{-st_1}. \tag{16.2.3}$$

The limiting version of the foregoing as $t_1 \to 0$ yields the transform of the derivative,

$$L(u'(t)) = sL(u) - u(0), \tag{16.2.4}$$

a relation which can be iterated to express $L(u^{(n)}(t))$ in terms of $L(u)$.

Finally, as Borel indicated, the Laplace transform unravels convolutions,

$$L\left(\int_0^t u(t_1)\, v(t - t_1)\, dt_1\right) = L(u)\, L(v). \tag{16.2.5}$$

The foregoing results hold under very light restrictions. See the references cited at the end of the chapter for rigorous derivations.

16.3. Simplifying Properties of the Laplace Transform

Let u satisfy the equation $T(u) = 0$. Using the foregoing properties, we find that for large classes of equations

$$L(T(u)) = T_1(L(u)) = 0, \tag{16.3.1}$$

where T_1 is a far more tractable transformation than T. This means that $L(u)$ is readily determined, often explicitly, in cases where u itself is not easy to obtain directly. We give some examples below.

The original problem of solving $T(u) = 0$ has thus been transformed into that of finding u given $L(u)$, which is to say, that of the inversion of the Laplace transform. We are particularly interested here in numerical inversion.

16.4. Examples

Before discussing various procedures for analytic and computational inversion of the Laplace transform, let us give some of the familiar examples of its use.

Consider first a linear differential equation with constant coefficients,

$$u'' + a_1 u' + a_2 u = g(t), \qquad u(0) = c_1, \quad u'(0) = c_2. \tag{16.4.1}$$

We have

$$L(u'' + a_1 u' + a_2 u) = L(g), \tag{16.4.2}$$

whence, using (16.2.4) twice,

$$L(u) = L(g) + \frac{(sc_1 + c_2 + a_1 c_2)}{(s^2 + a_1 s + a_2)}. \tag{16.4.3}$$

Consider next the renewal equation

$$u(t) = g(t) + \int_0^t k(t - s)\, u(s)\, ds. \tag{16.4.4}$$

Using (16.2.5), we obtain the relation

$$L(u) = \frac{L(g)}{1 - L(k)} \tag{16.4.5}$$

for $L(k) \neq 1$.

Finally, consider the linear partial differential equation

$$\begin{aligned} k(x)\, u_t &= u_{xx}, && 0 < x < 1, \quad t > 0, \\ u(x, 0) &= g(x), && 0 < x < 1, \\ u(0, t) &= u(1, t) = 0, && t > 0, \end{aligned} \tag{16.4.6}$$

an equation encountered in the theory of heat conduction. Then, proceeding formally,

$$sk(x)\, L(u) - k(x)\, g(x) = L(u)_{xx}, \qquad L(u)_{x=0} = 0 = L(u)_{x=1}, \tag{16.4.7}$$

an ordinary differential equation for $L(u)$, subject to a two-point boundary condition.

Rigorous derivations of the transformed equations may be found in the references cited at the end of the chapter. Here we are interested solely in illustrating the potential of the method.

Observe that in all cases the Laplace transform serves to lower the transcendence level of the equation. An ordinary differential equation and an integral equation are converted to algebraic equations, while a partial differential equation is transformed into an ordinary differential equation.

Exercise

1. Find $L(u)$ explicitly in the case where $u(t)$ satisfies the differential-difference equation $u'(t) = au(t) + bu(t-1)$, $t \geqslant 1$, $u(t) = g(t)$, $0 \leqslant t \leqslant 1$.

16.5. Complex Inversion Formula

Consider the expression

$$v(s) = L(u) = \int_0^\infty e^{-st} u(t)\, dt, \tag{16.5.1}$$

as a function of a complex variable s, where the integral is taken to be absolutely convergent for $\text{Re}(s) \geqslant s_0$. Then we have classically the fundamental inversion formula

$$u(t) = \frac{1}{2\pi i} \int_C v(s)\, e^{st}\, ds \tag{16.5.2}$$

where C is a suitably chosen contour. Often it is chosen to be a line,

$$s = s_1 + i\tau, \qquad -\infty < \tau < \infty, \quad s_1 > s_0 . \tag{16.5.3}$$

Starting with (16.5.2), the theory of residues can be profitably employed to yield many elegant representations of $u(t)$ which can be used for both qualitative and quantitative purposes. This requires a detailed knowledge of $v(s)$ as a function of a complex variable, in particular, the location of singularities of $v(s)$ and a knowledge of their nature. Sometimes this is easily done, sometimes not.

16.6. Abelian and Tauberian Results

When this determination is difficult, we often restrict our attention to $v(s)$ as a function of a real variable. It is clear that the behavior of $v(s)$ as $s \to \infty$ is governed by the behavior of $u(t)$ for t in the vicinity of zero; similarly, the behavior of $u(t)$ in the neighborhood of $t = \infty$ determines the form of $v(s)$ as $s \to 0$. These are results of Abelian type.

Conversely, under some mild conditions on $u(t)$, such as nonnegativity, one can show that the behavior of $u(t)$ as $t \to \infty$ is determined by the behavior of $v(s)$ as $s \to 0$. These are results of Tauberian type, more precisely of Hardy–Littlewood type. Thus a typical result is: if

$$\text{(a)}\quad u(t) \geqslant 0$$
$$\text{(b)}\quad \int_0^\infty u(t)\, e^{-st}\, dt \sim 1/s \qquad \text{as} \quad s \to 0, \tag{16.6.1}$$

then

$$\text{(c)}\quad \int_0^T u(t)\, dt \sim T \qquad \text{as} \quad T \to \infty.$$

Many other results of this nature are available.

What is quite interesting about this type of result is that it is an example of selective calculation. In favorable cases, given an equation for $u(t)$ of initial value type, we can employ the Laplace transform to ascertain the behavior of $u(t)$ as $t \to \infty$ without the necessity of calculating intermediate values in a time-consuming and frequently difficult step-by-step process.

16.7. Interpolation Technique

In some cases we can combine the behavior of $u(t)$ at $t = 0$ with the behavior at $t = \infty$ and some known structural properties of $u(t)$, such as monotonicity, convexity or concavity, to obtain a quick approximation to $u(t)$ which may be sufficient for immediate needs, or which may fruitfully be used in connection with a different method which is more precise but which requires a reasonable initial approximation.

16.8. Selective Calculation

Let us continue on this theme of extracting the desired amount of data from a relatively few calculations. Sometimes the values of $v(s)$ are readily obtained, as in the first and second examples above. On the other hand, sometimes the determination of each value of $v(s)$ requires a nontrivial calculation. One example of this is the heat equation

$$k(x)\, u_t = u_{xx}, \qquad u(x, 0) = g(x), \quad u(0, t) = u(1, t) = 0, \tag{16.8.1}$$

where the transformed equation is

$$g(x) + sk(x)v = v_{xx}, \qquad v(0) = v(1) = 0. \tag{16.8.2}$$

The situation is still more complex in the two-dimensional case,

$$k(x, y)\, u_t = u_{xx} + u_{yy}, \qquad u(x, y, 0) = g(x, y), \quad u(x, y, t) = 0, \, (x, y) \in B. \tag{16.8.3}$$

Here each value of s requires the solution of a potential equation.

$$g(x, y) - sk(x, y)v = v_{xx} + v_{yy}, \qquad v = 0, \quad (x, y) \in B, \tag{16.8.4}$$

a formidable problem in its own right.

Consequently, there is considerable interest in the question of using selected s-values of $v(s)$ to determine selected t-values of $u(t)$. Combined with Tauberian results, this would provide a powerful approach to the qualitative and quantitative study of functional equations.

16.9. Impossibility

An initial complication is that the task is impossible! To illustrate this, consider the well-known formula

$$\int_0^\infty \sin at \; e^{-st}\, dt = \frac{a}{a^2 + s^2}, \qquad a > 0, \quad s > 0. \tag{16.9.1}$$

For $s \geqslant 0$, $a \geqslant 0$, we have

$$\frac{a}{a^2 + s^2} \leqslant \frac{1}{a}. \tag{16.9.2}$$

Hence by choosing a arbitrarily large, we can make the Laplace transform of $\sin at$ arbitrarily small. Nonetheless, the function $\sin at$ oscillates between ± 1, a high-frequency "chatter."

It follows that an arbitrarily small change in the numerical value of $L(u)$ can produce arbitrarily large changes in the value of $u(t)$. Consequently, the Laplace inverse is an unstable operator.

At first sight, this is rather discouraging. What it means mathematically is that the original problem of numerical inversion is improperly posed. In order to convert it into a precise problem, we must add some additional information concerning the nature of $u(t)$, information which will rule out high frequency oscillations, spike functions and so on. In our case we shall restrict our attention to "smooth" functions, suitably defined below.

This type of consideration introduces a new, important and interesting class of mathematical problems, to wit: Given an equation $T(u) = 0$, how do we use information we may possess concerning the structure of the solution to aid in the discovery of further properties and particularly in the improved calculation of numerical values. We met this type of question briefly in Chapter 4, Volume I, in connection with the Emden–Fowler–Fermi–Thomas equation where we restricted our attention to "physical" solutions, i.e., those that exist for sufficiently large t. We shall meet it again in connection with the solution of ill-conditioned systems of linear equations.

Equations associated with physical processes always come with more information than explicitly stated in the equation. So to speak, they are imbedded in a scientific culture. It taxes the skill of the analyst to utilize this additional information in a systematic way. We shall give some examples below in connection with Tychonov regularization.

16.10. Quadrature Techniques

By a quadrature technique we shall mean an approximate formula of the type

$$\int_0^1 u(t)\,dt \cong \sum_{i=1}^{N} w_i u(t_i), \tag{16.10.1}$$

when the w_i and t_i are suitable chosen parameters, dependent on N, but not upon the function $u(t)$. The $2N$ parameters can clearly be chosen in many ways. For example, we can use the simple trapezoidal rule. The essential point is that different formulas require different numbers of quadrature points to ensure the same accuracy. If, as in many cases, the value at each quadrature point is obtained as a result of an intricate and time-consuming calculation, then we want to use as few quadrature points as possible.

The formula we shall employ is called Gaussian quadrature. As the exercises below show, the w_i (called the *weights*), and the t_i, the *nodes*, are uniquely determined by supposing that the formula is exact for polynomials of degree $2N - 1$ or less. It turns out that the t_i are the zeroes of the shifted Legendre polynomial of degree N, and the w_i are the Christoffel numbers, also given in terms of these polynomials.

Exercises

1. Show that the $n + 1$ moment conditions $\int_{-1}^{1} r^m p_n(r)\,dr = 0$, $m = 0, 1, \ldots, n - 1$, $\int_{-1}^{1} r^n p_n(r)\,dr = 1$ uniquely determine a polynomial of degree n. We use the interval $[-1, 1]$ here to make contact

with the Legendre polynomials. (Subsequently, we shall use the interval [0, 1] and the shifted Legendre polynomials.)

2. Consider the nth degree polynomial $\pi_n(r) = (d/dr)^n (r^2 - 1)^n$. Show that it satisfies the foregoing moment equations. Introduce the Legendre polynomials

$$p_n(r) = \frac{1}{2^n n!} \left(\frac{d}{dr}\right)^n (r^2 - 1)^n \qquad \text{(Rodriques' formula).}$$

3. Show that

$$\int_{-1}^{1} p_n(r)^2 \, dr = 2/(2n - 1), \qquad \int_{-1}^{1} p_m(r) \, p_n(r) \, dr = 0, \qquad m \neq n.$$

4. Show that we have the recurrence relations

$$(n + 1) \, p_{n+1}(r) - (2n + 1) \, r p_n(r) + n p_{n-1}(r) = 0, \qquad n = 1, 2,...$$

$$(r^2 - 1) \, p_n{'}(r) = n p_n(r) - n p_{n-1}(r), \qquad n = 1, 2,...,$$

with

$$p_0 = 1, \, p_1 = r, \qquad p_2 = \frac{(3r^2 - 1)}{2}, \qquad p_3 = \frac{5r^3 - 3r}{2}.$$

These results are important for computational purposes.

5. Show that $p_n(r)$ satisfies the linear differential equation

$$(1 - r^2) \frac{d^2u}{dr^2} - 2r \frac{du}{dr} + n(n + 1)u = 0.$$

6. Show from Rodriques' formula, or by use of orthogonality, that all the zeros of $p_n(r)$ are real and lie in the interval [—1, 1].

7. Show that the quadrature points are the zeroes of $p_n(r)$. (Hint: Consider the test polynomials $r^k p_n(r)$).

8. Determine the weights by using the test polynomial

$$f(r) = p_n(r)/(r - r_i)p'_n(r_i).$$

9. The foregoing establishes the necessity of the choices of the quadrature points and their weights; establish the sufficiency.

10. Establish the Darboux–Christoffel identity

$$p_0(x)\,p_0(y) + p_1(x)\,p_1(y) + \cdots + p_N(x)\,p_N(y)$$
$$= \frac{k_N}{k_{N+1}}\left[\frac{p_{N+1}(x)\,p_{N+1}(y) - p_N(x)\,p_N(y)}{x - y}\right]$$

where k_N is the coefficient of x^N in $p_N(x)$. (For the foregoing see

R. Bellman, R. Kalaba, and Jo Ann Lockett, *Numerical Inversion of the Laplace Transform*, American Elsevier, New York, 1966, Chapter 2,

and a number of references cited at the end of the chapter.

16.11. Numerical Inversion of the Laplace Transform

There are many existing methods for numerical inversion. This is to be expected, since we know in advance that no single method can work uniformly well. We shall present one which has worked out well in a number of cases, and is quite easy to use.

Introduce a new variable of integration, $x = e^{-t}$. Then the integral equation $L(u) = v(s)$ becomes

$$\int_0^1 x^{s-1}u(-\log x)\,dx = v(s). \tag{16.11.1}$$

Writing $g(x) = u(-\log x)$, we may consider then that we begin with the deceptively simple integral equation

$$\int_0^1 x^{s-1}g(x)\,dx = v(s). \tag{16.11.2}$$

Applying the quadrature formula of (16.10.1), this leads to the approximate relation

$$\sum_{i=1}^{N} w_i x_i^{s-1} g(x_i) = v(s). \tag{16.11.3}$$

If we now let s assume N different values, say $s = 1, 2, \ldots, N$, what results is a linear system of N equations for the N unknowns, $g(x_i)$, $i = 1, 2, \ldots, N$,

$$\sum_{i=1}^{N} w_i x_i^{k} g(x_i) = v(k+1), \qquad k = 0, 1, \ldots, N-1. \tag{16.11.4}$$

Presumably, we can now employ a standard program for solving linear algebraic equations, say Gaussian elimination combined with a pivoting policy. We can anticipate however, serious difficulties. The instability of the Laplace inverse should manifest itself in the ill-condition of the matrix (x_i^k), and so it does, as we shall see in a moment. It is clear that we can take $w_i g(x_i)$ as the fundamental variable so that the matrix of interest is (x_i^k), a Vandermonde matrix.

16.12. Explicit Inverse of (x_i^k)

What greatly improves the efficacy of the methods sketched above is the fact that we can readily determine the inverse of the Vandermonde matrix (x_i^k) in this case. Let us employ for this a classical device due to Jacobi.

Consider the system of equations

$$\sum_{i=1}^{N} x_i^k y_i = a_k, \qquad k = 0, 1, 2, \ldots, N-1. \tag{16.12.1}$$

Multiply the kth equation by q_k, a parameter as yet unspecified, and then add. The result is

$$\sum_{i=1}^{N} y_i \left(\sum_{k=0}^{N-1} q_k x_i^k \right) = \sum_{k=0}^{N-1} a_k q_k . \tag{16.12.2}$$

Hence, setting

$$f(x) = \sum_{k=0}^{N-1} q_k x^k, \tag{16.12.3}$$

we may write

$$\sum_{i=1}^{N} y_i f(x_i) = \sum_{k=0}^{N-1} a_k q_k . \tag{16.12.4}$$

The polynomial $f(x)$ of degree $N-1$ will be chosen expeditiously in a moment.

Let us ask that $f(x)$ be such that

$$f(x_i) = 0, \qquad f(x_j) = 1, \qquad i \neq j, \tag{16.12.5}$$

an orthogonality condition. With the $q_k \equiv q_{kj}$ determined in this fashion, we have

$$y_j = \sum_{k=0}^{N-1} a_k q_{kj} . \tag{16.12.6}$$

Let us do this for each j. Then (q_{kj}) is the inverse of the matrix (x_i^k). This method will clearly work for any Vandermonde matrix. Unfortunately, the determination of the required test polynomials, via say the Lagrange interpolation formula, and then of the q_{kj}, can introduce a great deal of numerical inaccuracy which may destroy the efficacy of the method.

In the present case, the fact that the x_i are the zeroes of the shifted Legendre polynomials p_N^* means that the Lagrange interpolation formula assumes a particularly simple form, namely

$$f_j(x) = \frac{p_N^*(x)}{(x - x_j)\, p_N'^*(x_j)}. \tag{16.12.7}$$

The determination of the q_{kj} is now tedious, but routine.

Exercises

1. Show that the coefficients in the $p_N^*(x)$ are integers.
2. Discuss how this information can be used to obtain the coefficients in $f_j(x)$ to arbitrary accuracy.

16.13. Example of Ill-conditioning

To illustrate what we mean by ill-conditioning, let us give the numerical values of some of the elements of the inverse of (x_i^k).

For $N = 7$, we have

2.4321935531218214*E*	1
−4.0620454164725030*E*	2
2.4240160528104807*E*	3
−6.8911106957307754*E*	3
1.0105700100432974*E*	4
−7.3801424274613136*E*	3
2.1240546327832863*E*	3

The integer following the values denotes the power of 10 by which the value is to be multiplied.

For $N = 9$, we have

3.9001694289454228*E*	1
−1.0602788886162712*E*	3
1.0622421003921418*E*	4
−5.3508791074809902*E*	4
1.5253238835016151*E*	5
−2.5700314282481323*E*	5
2.5355738385850332*E*	5
−1.3536662111862863*E*	5
3.0188269946893946*E*	4

For $N = 11$, we have

$$\begin{array}{rl}
5.7135446067451098E & 1 \\
-2.2931949503718826E & 3 \\
3.4449271098965383E & 4 \\
-2.6691108244372087E & 5 \\
1.2171732421675169E & 6 \\
-3.4859840680927631E & 6 \\
6.4479931659584596E & 6 \\
-7.6941942262124311E & 6 \\
5.7193040855846683E & 6 \\
-\ .40834\ 5386946350E & 6 \\
4.3874883894089764E & 5
\end{array}$$

The magnitude of the coefficients shows the origin of the instability.

16.14. Nonlinear Equations—I

Consider next the applicability of these techniques to nonlinear equations. As a simple example, consider the equation

$$u' = -u - u^2, \qquad u(0) = c, \quad |c| \ll 1. \tag{16.14.1}$$

We have

$$sL(u) - c = -L(u) - L(u^2), \qquad L(u) = \frac{c}{1+s} - \frac{L(u^2)}{1+s}. \tag{16.14.2}$$

Let the initial approximation be determined by

$$L(u_0) = \frac{c}{1+s}, \qquad u_0 = ce^{-t}, \tag{16.14.3}$$

and then determine successive approximations to the solution of (16.14.2) by means of

$$L(u_{n+1}) = \frac{c}{1+s} - \frac{L(u_n^2)}{1+s}. \tag{16.14.4}$$

An interesting point is that $L(u_n^2)$ is readily evaluated once u_n is known, using the available weights and quadrature points; namely

$$L(u_n^2) = \sum_{i=1}^{N} w_i u_n(t_i)^2. \tag{16.14.5}$$

Exercises

1. Under what conditions on $|c|$ does the method converge?

16.15. Nonlinear Equations—II

There is even an advantage in carrying out this procedure for ordinary differential equations, since it offers a procedure for selective calculation, and a way of avoiding "stiffness." Consider, however, the nonlinear differential-difference equation

$$\begin{aligned} u'(t) &= au(t) + bu(t-1) + u^2(t), \qquad && t \geqslant 1, \\ u(t) &= g(t), && 0 \leqslant t \leqslant 1. \end{aligned} \tag{16.15.1}$$

We can use the method outlined in the preceding section with about the same degree of effort as that required for an ordinary differential equation. When contrasted with the usual methods for computing the solution of a differential-difference equation, the proposed procedure possesses advantages—when it works.

Similarly, we can treat a nonlinear heat equation

$$\begin{aligned} k(x)\, u_t &= u_{xx} + g(u), \qquad && t > 0, \quad 0 < x < 1, \\ u(x, 0) &= h(x), && 0 < x < 1, \\ u(0, t) &= u(1, t) = 0, \end{aligned} \tag{16.15.2}$$

again with a substantial reduction of effort.

Exercises

1. Discuss the convergence of this method for both equations.

16.16. Quasilinearization

A significant drawback, nonetheless, to the foregoing method is the geometric convergence of the sequence $\{u_n\}$, when convergence occurs. It is not just a matter of time that is of moment here. It is a question of gambling at each stage on controlling the known ill-conditioning associated with each matrix inversion. Let us then turn to the use of quasilinearization where the convergence is quadratic—when it occurs. This could mean fewer stages.

Consider the simple case of the differential equation. Write,

$$L(u_{n+1}) = \frac{c}{1+s} - \frac{L(2u_n u_{n+1} - u_n^2)}{1+s}, \tag{16.16.1}$$

in place of (16.14.4). A difficulty, of course, is that when we use quadrature techniques on (16.16.1) we face the problem of inverting a different matrix at each stage due to the presence of the term $u_n u_{n+1}$. To circumvent this obstacle, we turn to Tychonov regularization, discussed in Sec. 2.27, Volume I.

Exercises

1. Consider the following method of successive approximations. Let x_1 be an initial approximation to the solution of $Ax = b$. Let x_2 be obtained as the value which minimizes

$$(Ax - b, Ax - b) + \lambda(x - x_1, x - x_1)$$

   and so on. Show that

$$x_{n+1} = (AA' + \lambda I)^{-1} A'b + \lambda(A'A + \lambda I)^{-1}\lambda_n .$$

2. Show that for $\lambda > 0$ the sequence $\{x_n\}$ converges and that the convergence is geometric, assuming, as we do, that A is nonsingular.

3. Show that the limit as $\lambda \to 0$ is the solution of $Ax = b$.

4. How does the rate of convergence depend on λ, and how does this fact affect the choice of λ?

5. How might one obtain an initial approximation? (Hint: Consider the use of small values of N first.)

6. Show that $x(\lambda) = x_{\min} + \lambda y_1 + \lambda^2 y_2 + \cdots$ for $|\lambda| \ll 1$.

7. Can one use this, plus the technique of deferred passage to the limit, to obtain a reasonable initial approximation?

8. Consider the differential equation $dx/dt = Ax - b$, $x_0 = c$. When does $x(t)$ converge as $t \to \infty$ to the solution of $Ax = b$?

9. Consider the equation $dx/dt = -A^T Ax + A^T b$, $x_0 = c$. Does the solution of this equation converge to the solution of $Ax = b$?

10. Consider the equation $dx/dt = g(Ax - b)$, $x(0) = c$, where $g(0) = 0$, and we want $\lim x(t)$ as $t \to \infty$. What choices of g are efficient? Are there any $g(x)$ where $g(0) = 0$, which are better than that in Exercise 9?

16.17. Tychonov Regularization

The problem we want to examine is that of solving $Ax = b$ where A is ill-conditioned. The temptation, of course, is to use a mean-square approach. In place of solving the original equation $Ax = b$, we tackle the problem of minimizing the function $(Ax - b, Ax - b)$. This leads to the equation $A'Ax = A'b$. There are several difficulties associated with this. In the first place, the matrix $A'A$ arising in this fashion may be more ill-conditioned than A. In the second place, because of the fact that A is ill-conditioned $\| Ax - b \| \leqslant \epsilon$ is a very light restriction on the range of x. It does not necessarily pin x down in any significant fashion.

It turns out, however, that a slight modification of the meansquare method yields a powerful procedure. We turn to the theme that we always possess more information concerning the nature of the desired solution than that expressed by an equation, in this case $Ax = b$. The challenge is always how to utilize these additional facts.

In this case we proceed as follows. Introduce the new function

$$(Ax - b, Ax - b) + \varphi(x) \tag{16.17.1}$$

where $\varphi(x)$ is chosen to take advantage of known properties of x. Thus, for example, we may know on the basis of earlier work that a reasonably good approximation to x is c. This leads to the choice

$$\varphi(x) = \lambda(x - c, x - c), \tag{16.17.2}$$

where λ is a parameter which must be adroitly chosen. If it is taken too large, the minimum value of the expression in (16.17.1) will remain close to c; if it is too small, then the a priori information $x \cong c$ will have no effect.

The choice of λ thus becomes an experimental matter, reinforcing our earlier point that computing must be viewed as a stochastic control process, indeed as an adaptive control process. What we learn at each stage of a computation is to be used to guide the subsequent stages.

The minimization of (16.17.1) where $\varphi(x)$ is as in (16.17.2) leads to the linear system

$$(A'A + \lambda I)x = A'b + \lambda c. \tag{16.17.3}$$

Even a small value of λ yields a matrix $A'A + \lambda I$ which substantially improves the ill-conditioning of $A'A$.

16.18. Self-consistent Methods

We can also make good use of knowledge of the structure of the components of x. For example, we may know from the origin of the problem that the components, $x_1, x_2, \ldots x_N$, vary smoothly as a function of the index; for example, monotonically. This suggests that we consider the minimization of

$$(Ax - b, Ax - b) + \lambda[(x_1 - x_2)^2 + (x_2 - x_3)^2 + \cdots + (x_{N-1} - x_N)^2] \tag{16.18.1}$$

or of

$$(Ax - b, Ax - b) + \lambda[(x_1 + x_3 - 2x_2)^2 + (x_2 + x_4 - 2x_3)^2 + \cdots + (x_{N-2} + x_N - 2x_{N-1})^2]. \tag{16.18.2}$$

This minimization can be conveniently carried out by means of dynamic programming as we indicate in the exercise below, or by using quadratic programming techniques if, for example, we impose the constraints $x_1 \leqslant x_2 \leqslant \cdots \leqslant x_N$.

Exercises

1. Consider the sequence of functions defined by

$$f_k(z, c) = \min_{x_n} \left[\sum_{i=1}^{M} \left(\sum_{j=k}^{N} a_{ij} x_j - z_i \right)^2 + \lambda[(x_N - x_{N-1})^2 + \cdots + (x_k - c)^2] \right].$$

2. Show that $f_N(z, c) = \min_{x_N} [\sum_{i=1}^{M} (a_{iN} x_N - z_i)^2 + \lambda(x_N - c)^2]$ and that

$$f_k(z, c) = \min_{x_k} [\lambda(x_k - c)^2 + f_{k+1}(z - a^k x_k, x_k)]$$

where

$$a^k = \begin{pmatrix} a_{1k} \\ a_{2k} \\ \vdots \\ a_{Mk} \end{pmatrix}.$$

3. Using the fact that $f_k(z, c) = (z, Q_k z) + 2(z, p_k)c + r_k c^2$ find recurrence relations for the matrix Q_k, the vector p_k and the scalar r_k.

4. Replace M by N, c by x_1, z_1 by $b_i - a_{i1} x_1$ and minimize the resultant quadratic polynomial over x_1 to solve the original problem.

5. Apply a similar procedure to the minimization of the Selberg quadratic form

$$S_N(x) = \sum_{k=1}^{N} \left(\sum_{v | a_k} x_v \right)^2,$$

where the notation $v \mid a_k$ means that the positive integer v divides the positive integer a_k and $\{a_k\}$ is a given sequence. See

A. Selberg, "On an Elementary Method in the Theory of Primes," *Kgl Norske Videnskab Selskabs Forh*, Bd 19, 1947, p. 18.

R. Bellman, "Dynamic Programming and the Quadratic Form of Selberg," *J. Math. Anal. Appl.*, Vol. 19, 1966, pp. 30–32.

16.19. An Imbedding Technique

We can, of course, use the same quadrature approach to treat general integral equations of the form

$$u(x) = f(x) + \int_0^1 k(x, y)\, u(y)\, dy. \tag{16.19.1}$$

The principal difficulty, of course, will reside in the solution of the associated system of linear algebraic equations.

Let us then consider another method which has yielded good results in applications. We make the restriction that $k(x, y) \geqslant a > 0$ for $0 \leqslant x, y < 1$. Let $\varphi(x, t)$ be defined by the solution of the linear integral equation

$$\phi(x, t) = g(x) + \int_0^1 \left[\int_0^t e^{-(t-s)/k(x,y)} \phi(y, s)\, ds \right] dy. \tag{16.19.2}$$

It is easy to see formally that the equation for $\phi(x, t)$ becomes the equation for $u(x)$ as $t \to \infty$.

It is not difficult to establish that if (16.19.1) has a unique positive solution $u(x)$ then (16.19.2) has a unique positive solution which converges to $u(x)$ as $t \to \infty$. We leave this and the other statement as exercises.

It remains to determine an algorithm for the computational solution of (16.19.2). If we employ quadrature techniques, we convert (16.19.2) into a system of ordinary differential equations. Thus, we obtain the approximate relations

$$\phi(x_i, t) = g(x_i) + \sum_{j=1}^{N} w_j \int_0^t e^{-(t-s)/k(x_i, y_j)} \phi(y_j, s)\, ds, \tag{16.19.3}$$

where $\{x_i\}$ and $y_j\}$ are the same sequences.

Exercises

1. What is the corresponding system of linear differential equations?
2. Show that the convergence is exponential.
3. Show that $\phi(x, t) \leqslant u(x)$ for $t \geqslant 0$ and that the convergence is monotone as $t \to \infty$.

16.20. Nonlinear Summability

A significant drawback to the foregoing method and to gradient imbedding methods in general, is that long-term integration is apparently required. This is expensive in time, but, most of all, round-off error can pile up and destroy the efficacy of the method. Fortunately, by using a nonlinear summability method we can avoid this. The method was described in Sec. 5.19, Volume I. We shall invoke it again below. Observe that it is a variation of a by now familiar theme: The specific structure of a particular solution should be used to improve a general computational algorithm.

Exercise

1. Examine the use of the foregoing imbedding technique in connection with the integral equations

$$\int_0^\infty e^{-st}u(t)\,dt = v(s), \qquad \int_0^1 r^{s-1}g(r)\,dr = h(s).$$

16.21. Time-dependent Transport Process

As a first example of the use of the Laplace transform in a partial differential-integral equation, consider the equation

$$u_x + 2u_t/c = a + a\int_0^t u_t(x, r)\,u(x, t - r)\,dr \tag{16.21.1}$$

which arises in time-dependent neutron transport theory. The initial conditions in time and space are

$$u(0, t) = 0, \qquad u(x, 0) = 0. \tag{16.21.2}$$

From the standpoint of numerical solution, there are several obstacles. First of all, it is a partial differential equation, secondly it is nonlinear. But worst of all there is the convolution term which makes serious demands on rapid access storage and time.

Let us bypass all of these by using the Laplace transform. Let $f(x, s) = L(u)$ for $\mathrm{Re}(s)$ sufficiently large. Then f satisfies a familiar equation.

$$\frac{df}{dx} = f^2 - 2sf + 1, \qquad f(0, s) = 0, \tag{16.21.3}$$

a Riccati equation which in this case can be resolved analytically, namely

$$f(x, s) = \frac{\tan(x\sqrt{1 - s^2})}{\sqrt{1 - s^2} + s\tan(x\sqrt{1 - s^2})}. \tag{16.21.4}$$

Exercises

1. Show that as $t \to \infty$

$$\begin{aligned} u(x, t) &\sim \tan x, && x < \pi/2 \quad \text{(subcritical)} \\ u(x, t) &\sim t, && x = \pi/2 \quad \text{(critical)} \\ u(x, t) &\sim a(x)\, e^{\lambda(x)t}, && x > \pi/2 \quad \text{(supercritical)} \end{aligned}$$

for positive quantities $a(x)$ and $\lambda(x)$.

2. Determine these quantities numerically.
3. Examine the existence and the uniqueness of the solution of (16.21.1). See

   G. M. Wing, *Introduction to Transport Theory*, Wiley, New York, 1962.

16.22. Radiative Transfer

The problem of the reflection and transmission of parallel rays of light of uniform intensity incident upon a plane parallel slab has led to a great deal of important and interesting analysis and there still remain a number of unresolved physical and mathematical questions. A classical imbedding of the type described in Chapter 13 leads to a linear transport equation for the internal flux $N(x, \mu)$

$$\mu \frac{\partial N}{\partial x} + \sigma N = \frac{c}{2} \int_{-1}^{1} N(x, \mu')\, d\mu', \tag{16.22.1}$$

an equation of deceptive simplicity. Typical boundary conditions are

$$\begin{aligned} N(-b, \mu) &= g(\mu), \qquad \mu > 0, \\ N(+b, \mu) &= h(\mu), \qquad \mu < 0. \end{aligned} \tag{16.22.2}$$

The parameters μ, σ, c, all have physical significance explained in the references given at the end of the chapter.

Exercises

1. Show that (16.22.1) can be solved in terms of a Wiener–Hopf integral equation.

16.23. Radiative Transfer via Invariant Imbedding

It turns out that it is not easy, even with large computers and quadrature techniques to solve (16.22.1) computationally. Hence, a great deal of sophisticated analysis and invariant imbedding is invoked. Let us consider only the invariant imbedding approach here. If we introduce the function

$$R(v, u, x) = \text{the intensity of flux reflected from a plane-parallel slab at an angle } \cos^{-1} v \text{ due to an incident flux at angle } \cos^{-1} u, \tag{16.23.1}$$

simple arguments yield the nonlinear partial differential integral equation,

$$\begin{aligned} \frac{1}{\lambda(x)} \left[R_x + \left(\frac{1}{u} + \frac{1}{v} \right) R \right] &= \left[1 + \frac{1}{2} \int_0^1 R(v, u', x) \frac{du'}{u'} \right] \\ &\quad \times \left[1 + \frac{1}{2} \int_0^{'} R(v', u, x) \frac{dv'}{v'} \right], \end{aligned} \tag{16.23.2}$$

with the initial condition

$$R(v, u, 0) = 0. \tag{16.23.3}$$

To obtain a computational solution, we employ quadrature techniques. Write

$$R(v_i, u_j, x) = f_{ij}(x). \tag{16.23.4}$$

Then (16.23.2) leads to the system of ordinary nonlinear differential equations

$$\frac{1}{\lambda(x)}\left[\frac{d}{dx}f_{ij}(x)+\left(\frac{1}{u_i}+\frac{1}{v_j}\right)f_{ij}(x)\right]$$

$$=\left[1+\frac{1}{2}\sum_{k=1}^{N}\frac{w_k}{u_k}f_{ik}(x)\right]\left[1+\frac{1}{2}\sum_{k=1}^{N}\frac{w_k}{v_k}f_{kj}(x)\right], \tag{16.23.5}$$

$i, j = 1, 2,..., N$, with the initial conditions

$$f_{ij}(0) = 0. \tag{16.23.6}$$

Quite satisfactory results have been obtained for $N = 7$ and 9, leading to 49 and 81 simultaneous equations respectively. Even $N = 5$ gives a good quick approximation. These figures of 49 and 81 can be reduced by symmetry considerations, and even further by some analytic ingenuity as we shall see below.

Exercises

1. Show that $R(v, u, x)$ is symmetric in v and u.

2. Let $\phi(u, x) = 1 + \frac{1}{2}\int_0^1 R(w, u, x)\, dv/v$. Show that ϕ satisfies the integral equation

$$\phi(u, x) = 1 + \frac{1}{2}\int_0^1 \left\{\exp\left(-\left(\frac{1}{u}+\frac{1}{v}\right)x\right)\right.$$

$$\left.\int_0^x \exp\left(\left(\frac{1}{u}+\frac{1}{v}\right)y\right)\lambda(y)\,\phi(u, y)\,\phi(v, y)\,dy\right\}\frac{dv}{v}$$

$$= T(\phi).$$

3. Show that this equation possesses a unique solution for $x \geqslant 0$, $0 \leqslant u, v \leqslant 1$, that this solution is uniformly bounded in $0 \leqslant x < \infty$ provided that $|\lambda(x)| \leqslant 1$ for $x \geqslant 0$, and that it may be found as the limit of the successive approximations,

$$\phi_0(u, x) = 1, \qquad \phi_{n+1}(u, x) = T(\phi_n).$$

4. Consider the case where $\lambda(x) \equiv \lambda$ (homogeneous medium). Show that in this case $\phi(u, x)$ is monotonically increasing in x and that $\lim_{x\to\infty}\phi(u, x)$ exists.

5. Write $f(u) = \lim_{x\to\infty} \phi(u, x)$. Show that $f(u)$ satisfies the Ambarzumian–Chandrasekhar equation

$$f(u) = 1 + \frac{\lambda}{2}\int_0^1 \frac{uf(u)f(v)}{u+v}\,dv.$$

6. Discuss the existence and uniqueness of the solution of this equation from first principles when $0 < \lambda \leqslant 1$.

16.24. Time-dependent Case

A time-dependent version of the radiative transfer process leads to a truly formidable nonlinear partial differential integral equation

$$\begin{aligned} &g_x + (u^{-1} + v^{-1})(g_t + g) \\ &\quad = \lambda \Big[\frac{H(t)}{4v} + \frac{1}{2}\int_0^1 g(v, u', x, t)\frac{du'}{u} + \frac{1}{2v}\int_0^1 g(v', u, x, t)\,dv' \\ &\qquad + \int_0^t dt' \int_0^t g(v, u', x, t')\,dv' \int_0^1 g_t(v, u', x, t-t')\frac{du'}{u'}\Big], \end{aligned} \tag{16.24.1}$$

with $g(v, u, x, 0) = 0$. Here $\lambda = \lambda(x)$ and

$$\begin{aligned} H(t) &= 0, \qquad t < 0 \\ &= 1, \qquad t > 0, \end{aligned} \tag{16.24.2}$$

Nonetheless, it yields quite tamely to a combination of the Laplace transforms, quadrature and Laplace inversion. The transformed equation is, setting $f(v, u, x, s) = 4vL(g)$,

$$\begin{aligned} &f_x + (s+1)(u^{-1} + v^{-1})f \\ &\quad = \lambda \Big\{\frac{1}{s} + \frac{1}{2}\int_0^1 f(v, u', x, s)\frac{du'}{u'} + \frac{1}{2}\int_0^1 f(v', u, x, s)\frac{dv'}{v'} \\ &\qquad + \frac{s}{4}\int_0^1\int_0^1 f(v', u, x, s)f(v, u', x, s)\frac{dv'}{v'}\frac{du'}{u'}\Big\}, \end{aligned} \tag{16.24.3}$$

with $f(v, u, 0, s) = 0$. For each value of s this may be treated as outlined in Section 16.23. This is a nice example of the point we made before that an extensive calculation may be required to provide a particular value of $f(v, u, x, s)$.

Details will be found in the references cited.

16.25. Error Analysis

Let us now say a few words about how well we expect the numerical inversion to work. This will illustrate a point previously stressed that every computational procedure leads to a stability question, stability not of the original equation, but of the approximation used for numerical purposes.

Our aim is to solve the linear integral equation

$$\int_0^1 x^{s-1}g(x)\,dx = v(s), \tag{16.25.1}$$

We consider instead the approximate relation

$$\sum_{i=1}^{N} w_i x_i^{s-1} g(x_i) = v(s), \qquad s = 1, 2, \ldots, N. \tag{16.25.2}$$

We can write (16.25.1) in the form

$$\sum_{i=1}^{N} w_i x_i^{s-1} g(x_i) = v(s) + \left(\sum_{i=1}^{N} w_i x_i^{s-1} g(x_i) - \int_0^1 x^{s-1} g(x)\,dx\right). \tag{16.25.3}$$

Our computational procedure consists of neglecting the term in parenthesis.

Thus we see that we have made two tacit assumptions:

$$\text{(a)}\quad \sum_{i=1}^{N} w_i x_i^{k} g(x_i) - \int_0^1 x^k g(x)\,dx = \Delta_k \tag{16.25.4}$$

is small for $k = 0, 1, \ldots, N-1$

(b) The difference between the solution of (16.25.2) for $s = 1, 2, \ldots, N$ and that of $\sum_{i-1}^{N} w_i x_i^k\, g(x_i) = v(k+1) + \Delta_k$, $k = 0, 1, \ldots, N-1$ is small.

The first assumption is valid provided that $g(x)$ can be well approximated by a polynomial of degree $2N-1$, or less, which is to say $g(t) \cong \sum_{r=0}^{2N-1} a_r e^{-rt}$. The second can be validated using the explicit inverse of (x_i^k) once we have an estimate for Δ_k. Since (x_i^k) is an ill-conditioned matrix, a sufficient condition is that $|\Delta_k|$ be extremely small; cf. Section 16.13 where we give some examples of the values of the components in the inverse matrix for $N = 7, 9, 11$.

The remarkable fact is that although these estimates obtained in this way make us quite pessimistic about the value of the method, in practice it works very well. This hinges upon the choice of s-values, whether we choose $s = 1, 2, \ldots, N$ or $s = 10, 20, \ldots, 10N$. Particularly in dealing with ill-conditioned systems, or, more generally improperly posed problems, a certain amount of experimentation (trial and error), is required. Recall our point that computing must be regarded as an adaptive control process.

In practice, we use different values of N, say $N = 7, 9, 11$, with different degrees of accuracy in the determination of $v(k + 1)$ and compare the results. If they are self-consistent and agree with our mathematical and physical intuition, we accept them. This is the kind of procedure used in the numerical solution of ordinary differential equations, partial differential equations and so on. There the adjustable parameter is the grid size.

Nonetheless, the efficacy of the method is quite mysterious. It would bc nice to have an explanation.

16.26. Differential Quadrature

We can also use quadrature techniques to eliminate derivatives. Our aim is to replace a derivative $f'(x_i)$ by a linear combination of the values $f(x_j)$, $j = 1, 2, \ldots, N$,

$$f'(x_i) = \sum_{j=1}^{N} a_{ij} f(x_j). \tag{16.26.1}$$

Let us discuss two different procedures we can use to choose the coefficients. In the first place we can ask that the relation be valid for any polynomial $f(x)$ of degree less than or equal to $N - 1$. Let us employ the test function

$$f(x) = \frac{p_N{}^*(x)}{(x - x_k)\, p_N^{*\prime}(x_k)}. \tag{16.26.2}$$

Then

$$a_{ik} = \frac{p_N^{*\prime}(x_i)}{(x_i - x_k)\, p_N^{*\prime}(x_k)}, \qquad k \neq i$$

$$a_{kk} = \frac{p_N^{*v}(x_k)}{2p_N^{*\prime}(x\ )} = \frac{1 - 2x_k}{2(x_k{}^2 - x_k)}. \tag{16.26.3}$$

Secondly, we can assume that $f(x)$ is well approximated to by a spline and use the linear relation connecting $f'(x_i)$ and the $f(x_j)$ derived by the method in Sec. 12.28.

16.27. Application to Partial Differential Equations

Consider the nonlinear first-order partial differential equation

$$u_t = g(t, x, u, u_x), \qquad -\infty < x < \infty, \quad t > 0 \tag{16.27.1}$$

with the initial condition $u(0, x) = h(x)$. Write

$$u_x(t, x_i) = \sum_{j=1}^{N} a_{ij} u(t, x_j), \qquad i = 1, 2, \ldots, N. \tag{16.27.2}$$

Then (16.27.1) yields the set of N ordinary differential equations

$$\begin{aligned} u_t(t, x_i) &= g(t, x_i, u(t, x_i), \sum_{j=1}^{N} a_{ij} u(t, x_j)), \\ u(0, x_i) &= h(x_i), \qquad i = 1, 2, \ldots, N. \end{aligned} \tag{16.27.3}$$

Exercises

1. Test the method using the equation $u_t = uu_x$, $0 < x < 1$, $0 < t$, with $u(x, 0) = x/10$, and $u(x, 0) = (\sin \pi x)/10$.
2. Test the method with the system $u_t = uu_x + vu_y$, $v_t = uv_x + vv_y$. See

   R. Bellman, B. Kashef, and J. Casti, "Differential Quadrature: A Technique for the Rapid Solution of Nonlinear Partial Differential Equations," (to appear).
3. Consider the numerical stability of the method in connection with the linear equation $u_t = u_x$, $u(x, 0) = h(x)$.

16.28. Identification Problems

A problem of great scientific importance is one where it is necessary to use observations of the solution to determine parameters in the equation. One formulation is the following.

Given the equation

$$x' = g(x, a), \qquad x(0) = c, \tag{16.28.1}$$

where the parameter a and the initial condition c are unknown, determine the value, or values of a and c, such that $\{x(t_i)\}$, $i = 1, 2, \ldots, N$, is a prescribed set of vectors (observations).

The problem is a very difficult one and little has been done concerning existence or uniqueness. It is the basic problem in mathematical model-making.

16.29. Quasilinearization

One simple approach which has worked well in many fields is furnished by the theory of quasilinearization. Let us pose the problem in the following form: Determine a and c so as to minimize the expression

$$\sum_{i=1}^{N} \| x(t_i, a, c) - b_i \|^2 \tag{16.29.1}$$

where we write $x(t, a, c)$ as the solution of (16.28.1) and b_i as the value of $x(t_i)$.

Let a_0, c_0 be an initial approximation, leading to the value x_0 determined by $x_0' = g(x_0, a_0)$, $x_0(0) = c_0$. Write as the next approximation

$$x_1' = g(x_0, a_0) + J_1(x_1 - x_0) + J_2(a_1 - a_0), \qquad x_1(0) = c_1, \tag{16.29.2}$$

where J_1 and J_2 are the appropriate Jacobians. Then x_1 is a linear function of a_1 and c_1, which means that (16.29.1) becomes a quadratic variational problem which is readily resolved. This procedure yields a new triple (x_1, a_1, c_1) and the method continues.

Exercises

1. Do we always have convergence?
2. Is there convergence if a_1 and c_1 are sufficiently close to the minimizing values?
3. What are some disadvantages to the foregoing method?

16.30. Differential Quadrature

Another simple technique uses differential quadrature. Write as before

$$x'(t_i) = \sum_{j=1}^{M} a_{ij} x(t_j), \qquad i = 1, 2, \ldots, M \tag{16.30.1}$$

and replace the differential equation by

$$\sum_{j=1}^{N} a_{ij}x(t_j) = g(x(t_i), a), \qquad i = 1, 2,..., M, \tag{16.30.2}$$

an overdetermined system in general. Hence we replace (16.30.2) by the problem

$$\min_{a} \sum_{i=1}^{M} \left\| \sum_{j=1}^{N} a_{ij}x(t_j) - g(x(t_i), a) \right\|^2. \tag{16.30.3}$$

Here we are assuming that c is known.

In many important cases, $g(x, a)$ is linear in the components of a,

$$g(x, a) = \sum_{i=1}^{K} a_i g_i(x), \tag{16.30.4}$$

which makes (16.30.3) a quadratic minimization problem quite easily resolved.

Miscellaneous Exercises

1. Write $v(s) = \int_0^\infty u(t)e^{-st}\,dt$. Show that under appropriate assumptions, one has the Post–Widder inversion formula

$$u(t) = \lim_{k\to\infty} \left[\frac{(-1)^k}{k!}\left(\frac{k}{t}\right)^{k+1} v^{(k)}\left(\frac{k}{t}\right)\right].$$

2. Motivate this formula starting with the Parseval formula

$$v(s) = \int_0^\infty \frac{w(r)}{t + ir}\,dr,$$

where $w(r)$ is the Fourier transform of $u(t)$.

3. Using the formula

$$e^{-\sqrt{s}} = \int_0^\infty \frac{e^{-st}e^{-t/4}\,dt}{2\sqrt{\pi}\,t^{3/2}}$$

test the numerical validity of the formula for $k = 1, 2, 3$ for some representative t-values. See

R. Bellman, R. Kalaba, and J. Lockett, *Numerical Inversion of the Laplace Transform*, American Elsevier, New York, 1966, p. 16.

4. Show that the zeroes of $p_N^*(x)$, x_{iN}, interlace in [0, 1] and discuss how this fact can be used numerically.

5. Consider the Mellin transform, $M(u) = \int_0^\infty u(t)t^{s-1}\,dt$, $\mathrm{Re}(s) > 0$. Make the change of variable $t = r/(1-r)$ so that

$$M(u) = \int_0^1 g(r)\left(\frac{r}{1-r}\right)^{s-1} dr,$$

with $g(r) = u \mid (1-r)^2$.

6. Apply Gaussian quadrature to obtain the system of linear algebraic equations

$$f(k+1) = \sum_{i=1}^{N} w_i g(r_i)\left(\frac{r_i}{1-r_i}\right)^{k-1}, \qquad k = 1,\ldots, N.$$

7. Obtain an explicit solution of this system of equations. (See the book previously cited.)

8. Consider an approximation in the s-plane, $v(s) = \sum_{k=1}^{N} a_k/(s+\lambda_k)$, obtained by use of a Padé approximation. A detailed discussion of alternate inversion techniques with further references is given in

   T. L. Cost, "Approximate Laplace Transform Inversions in Visco–elastic Stress Analysis," *AIAA J.*, Vol. 2, 1964, pp. 2157–2166.

9. Consider the equation $u^{(4)} - 22u^{(3)} + 39u^{(2)} + 22u^{(1)} - 40u = 0$ subject to initial conditions $u(0) = 1$, $u'(0) = -1$. $u^{(2)}(0) = 1$, $u^{(3)}(0) = -1$. Use an Adams–Moulton integration method with grid size of 0.01 to integrate out to $t = 2$. What happens and why?

10. Show that the Laplace transform can be used to overcome this problem.

11. Obtain the Rodriques formula using the generating function $(1 - 2rt + t^2)^{-1/2}$ and the Lagrange expansion formula.

12. The equation $\mathrm{a}u_t = -u_y + k(v-u)$, $bv'(t) = k\int_0^1 u\,dy - kv$, $u(y, 0) = 0$, $u(0, t) = u_0$, $0 < t < t_1$, $= 0$, $t_1 < t \leqslant T$, $= u(1, t-T)$, $t > T$, arises in chemotherapy. Obtain the solution by means of the Laplace transform.

13. Solve the Abel integral equation using the Laplace transform,

$$v(t) = \int_0^t (t-t_1)^{a-1}\,u(t_1)\,dt_1\,.$$

14. Consider the minimization of $(Ax - b, Q(Ax - b)) + \lambda(x - c, x - c)$, $\lambda > 0$ where Q is a suitably chosen positive definite matrix and c is an initial approximation. Is there any advantage to a control process of the following type: Choose

$$Q = \begin{pmatrix} |(Ax_0 - b)_1| & & & 0 \\ & |(Ax_0 - b)_2| & & \\ & & \ddots & \\ 0 & & & |(Ax_0 - b)_N| \end{pmatrix},$$

where $(Ax_0 - b)_1$ is the first component of $Ax_0 - b$ and so on, c, the initial approximation becomes x_1 and λ is suitably modified? This is essentially a relaxation process. See

K. Levenberg, "A Method for the Solution of Certain Nonlinear Problems in Least Squares," *Quart. Appl. Math.*, Vol. 2, 1944, pp. 164-168.

15. Consider the following extension of the Steinhaus approximation procedure presented in Volume 1. Let $f(x, y) = 0$, $g(x, y) = 0$ be two simultaneous equations with a unique solution. Let (x_0, y_0) lie upon $f(x, y) = 0$ and let (x_1, y_1) be determined as the foot of the normal to $g(x, y) = 0$ from (x_0, y_0), i.e., (x_1, y_1) is the point on $g(x, y) = 0$ closest to (x_0, y_0). Repeat the process starting from (x_1, y_1). When do we have convergence? Does the distance from (x_n, y_n) to the common solution decrease using a suitable norm?

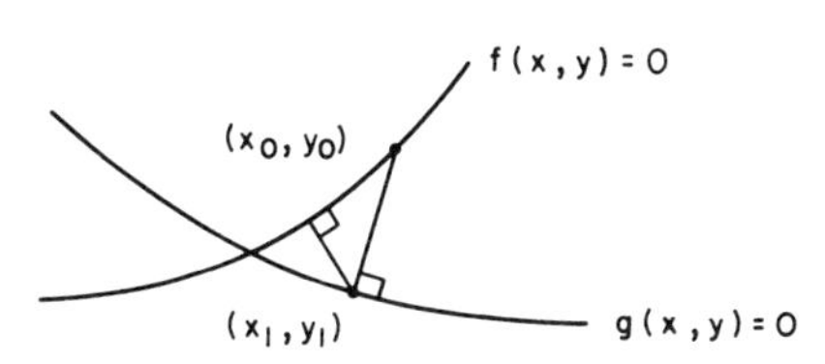

16. Let $L(u, v) = 0$, $M(u, v) = 0$ be two linear functional equations with a unique solution and (u_0, v_0) be a function pair satisfying $L(u, v) = 0$. Let (u_1, v_1) be the function pair minimizing the distance $\| u - u_0 \| + \| v - v_0 \|$ subject to $M(u, v) = 0$. Repeat the process. When do we have convergence? Is there a distance function which is decreased at each stage?

17. Consider the equations

$$\varphi(x) \int_a^b g(x, y)\, \psi(y)\, dy = h_1(x), \qquad \psi(y) \int_c^d g(x, y)\, \varphi(x)\, dx = h_2(y)$$

where $h_1(x)$ and $h_2(y)$ are given continuous functions and $g(x, y)$ is a probability density function, e.g.,
$g(x, y) = \exp[-(x - y)^2/\sigma^2] \sigma\sqrt{\pi}$, $\sigma > 0$ and discuss the possibility of positive solutions.

R. Fortet, "Resolution d'un Systeme d'equations de M. Schrödinger," *J. Math. Pures Appl.*, Vol. 9, 1940, pp. 83–105.

These equations arise in connection with some very interesting inverse problems in quantum mechanics. See

E. Schrödinger, Sur la Theorie Relativiste de l'electron, *Amer. Inst. Henri–Poincare*, Vol. 2, 1932, p. 300.

E. Schrödinger, Uber die Umkehrung der Naturgesctze, *Sitzung Phys.–Math. Klasse*, March, 1931, pp. 144–153.

In order to test both analytic approximations and computational techniques and to build up intuition, it is essential to possess classes of nonlinear partial differential equation with explicit analytic solutions. The exercises below indicate some of the ways in which this can be done.

18. Show that the Burgers' equation $u_t + uu_x = \epsilon u_{xx}$, can be solved in terms of the solution of the linear heat equation $\epsilon w_{xx} = w_t$ (Cole–Hopf).
(Hint: Set $u = -2\epsilon w_x/w$). Thus, the Burgers' equation may be regarded as a Riccati equation associated with the heat equation. See

E. Hopf, "The Partial Differential Equation $u_t + uu_x = \mu u_{xx}$," *Comm. Pure Appl. Math.*, Vol. 3, 1950, pp. 201–230.

19. Similarly, linearize

$$u_t + uu_x + vu_y = \epsilon(u_{xx} + u_{yy}),$$

$$v_t + uv_x + vv_y = \epsilon(v_{xx} + v_{yy}).$$

20. Consider the equation

$$\frac{\partial f}{\partial t} + y\frac{\partial f}{\partial x} = \frac{\partial^2 f}{\partial y^2},$$

with $f(x, y, 0) = g(x, y)$. Introduce the moments

$$\rho = \int_{-\infty}^{\infty} f(x, y, t)\, dy, \qquad \rho u = \int_{-\infty}^{\infty} yf(x, y, t)\, dy,$$

Assume that f and yf vanish at $y = \pm\infty$. Show that we have

$$\frac{\partial \rho}{\partial t} + \frac{\partial}{\partial x}(\rho u) = 0, \qquad \frac{\partial u}{\partial t} + u\frac{\partial u}{\partial x} + \frac{p_x}{\rho} = 0$$

See

A. H. Taub, "A Sampling Method for Solving the Equations of Compressible Flow in a Permeable Medium," *Proc. of First Midwestern Conference on Fluid Dynamics*, Edwards, Ann Arbor, Michigan, 1951, pp. 121–127.

21. How can $g(x, y)$ be chosen to ensure that an equation of state, $p = p(\rho)$, holds? How would one ensure that it holds approximately?

22. Begin with the equation $\partial f/\partial t + y(\partial f/\partial x) = \epsilon(\partial^2 f/\partial x^2)$. With the same definitions of ρ, u, and p as above, show that

$$\frac{\partial \rho}{\partial t} + \frac{\partial}{\partial x}(\rho u) = \epsilon\frac{\partial^2 \rho}{\partial x^2},$$

$$\frac{\partial u}{\partial t} + u\frac{\partial u}{\partial x} + \frac{p_x}{\rho} = \epsilon\frac{\partial^2 u}{\partial x^2} + 2\epsilon\frac{\partial u}{\partial x}\frac{\partial \rho}{\partial x}.$$

23. Obtain two-dimensional versions of the equations in Exercise 22. This is connected with extensive efforts to derive the equations of hydrodynamics from the Boltzmann equation, research of Hilbert, Enskog and Grad.

Bibliography and Comment

§16.1. A great deal of attention has been devoted to quadrature techniques. See

C. Lanczos, *Applied Analysis*, Prentice-Hall, Englewood Cliffs, New Jersey, 1956.

J. A. Shohat, and J. D. Tamarkin, *The Problem of Moments*, Amer. Math. Soc., Providence, Rhode Island, 1943.

J. Todd, (ed.), *Survey of Numerical Analysis*, McGraw-Hill, New York, 1962.

H. B. Keller, "Approximate Solutions of Transport Problems. II. Convergence and Applications of the Discrete Ordinate Method," *J. Soc. Indust. Appl. Math.*, Vol. 8, No. 1, March 1960, pp. 43–73.

H. B. Keller, "On the Pointwise Convergence of the Discrete-Ordinate Method," *J. Soc. Indust. Appl. Math.*, Vol. 8, 1960, pp. 560–567.

A significant extension is the finite element method of modern engineering.

The quadrature technique, like the finite difference technique, can be improved by the method of deferred correction. See

V. Pereyra, *Accelerating the Convergence of Discretization Algorithms*, Mathematics Research Center, Univ. Wisconsin, Madison, Wisconsin, Technical Summary Rep. 687, October 1966.

V. Pereyra, "On Improving an Approximate Solution of a Functional Equation by Deferred Corrections," *Numer. Math.*, Vol. 8, 1966, pp. 376–391.

§§16.2–16.3. Detailed discussions of the Laplace transform may be found in

G. Doetsch, *Handbuch der Laplace-Transformation*, 3 vols., Basel, 1950–1956.

D. V. Widder, *The Laplace Transform*, Princeton Univ. Press, Princeton, New Jersey, 1941.

R. Bellman, and K. L. Cooke, *Differential-Difference Equations*, Academic Press, New York, 1963.

§16.4. The renewal equation can be conveniently treated in some cases by differential approximation (see Chapter 6, Volume I) and in other cases by the discrete stage technique of Kendall (see Chapter 14).

§16.6. See the books cited for Secs. 16.2–16.3.

§16.8. See

E. Angel and R. Bellman, *Dynamic Programming and Partial Differential Equations*, Academic Press, New York, 1972.

for illustrations of the successful use of this method on the heat equation.

§16.10. The properties of Legendre polynomials are discussed extensively in

E. T. Whittaker, and G. N. Watson, *Modern Analysis*, Cambridge Univ. Press, London and New York, 1935.

Szego, G., *Orthogonal Polynomials*, Amer. Math. Soc., Providence, Rhode Island, 1939.

§16.11. We are following the method first presented in

R. Bellman, R. Kalaba, and J. Lockett, *Numerical Inversion of the Laplace Transform*, Amer. Elsevier, New York, 1966.

Many numerical examples may be found there.

§16.15. Many interesting and difficult questions arise in connection with the computational solution of differential-difference and integro–differential equations; see

K. L. Cooke and S. E. List, "The Numerical Solution of Integro-Differential Equations with Retardation," (to appear).

§16.17. See the discussion in the book

R. Bellman, R. Kalaba, and J. Lockett, *Numerical Inversion of the Laplace Transform*, Amer. Elsevier, New York, 1966.

and the books

A. Lavrientiev, *Some Improperly Posed Problems of Mathematical Physics*, Springer–Verlag, Berlin and New York, 1967.

R. Lattes and J. Lions, *The Method of Quasi–Reversibility: Applications to Partial Differential Equations*, (translated by R. Bellman), Amer. Elsevier, New York, 1969.

for a general discussion of improperly posed problems.

See the papers

A. Tikhonov, "Regularization of Incorrectly Posed Problems," *Soviet Math. Dokl.* Vol. 4, (163).

D. L. Philips, "A Technique for the Numerical Solution of Certain Integral Equations of the First Kind," *J. Assoc. Comput. Math.*, Vol. 9, 1962, pp. 84–97.

S. Twomey, "On the Numerical Solution of Fredholm Integral Equations of the First Kind ...," *J. Assoc. Comput. Mach.*, Vol. 10, 1963, pp. 97–101.

K. Levenberg, "A Method for the Solution of Certain Nonlinear Problems in Least Squares," *Quart. Appl. Math.*, Vol. 2, 1944, pp. 164–168.

§16.19. See

R. Bellman, "A New Technique for the Numerical Solution of Fredholm Integral Equations," *Comput. Rev.*, Vol. 3, 1968, pp. 131–138.

For applications to some astrophysical problems, see

J. Gruschinske and S. Ueno., "Bellman's New Approach to the Numerical Solution of the Auxiliary Equation in Spherical Medium," *Publ. Astronom Soc. Japan*, Vol. 22, 1970, p. 365.

J. Gruschinske and S. Ueno., "Bellman's New Approach to the Numerical Solution of Fredholm Integral Equations with Positive Kernels," *J. Quant. Spectrosc. and Radiat. Transfer*, Vol. 11, 1971, pp. 641–646.

§16.20. For an application of this extrapolation technique, see

R. Bellman and R. Kalaba, "A Note on Nonlinear Summability Techniques in Invariant Imbedding," *J. Math. Anal. Appl.*, Vol. 6, 1963, pp. 465–472.

For a detailed discussion and numerous applications, see

D. Shanks, "Nonlinear Transformations of Divergent and Slowly Convergent Sequences," *J. Mathematical Phys.*, Vol. 34, 1955, pp. 1–42.

§16.21. For a detailed discussion of this equation see the book

G. M. Wing, *An Introduction to Transport Theory*, Wiley, New York, 1962.

§16.22. See the book by Wing cited above and

S. Chandrasekhar, *Radiative Transfer*, Oxford Univ. Press, London and New York, 1950.

§16.23. See

R. Bellman, R. Kalaba, and M. Prestrud, *Invariant Imbedding and Radiative Transfer in Slabs of Finite Thickness*, Amer. Elsevier, New York, 1963.

This direct approach can be considerably improved by using some analytic considerations of Chandrasekhar; see the book cited above.

For the Ambarzumian–Chandrasekhar equation, see

C. Fox, *Trans. Amer. Math. Soc.*, Vol. 99, 1961, pp. 285–291.

V. A. Kakicev and V. S. Rogozin, "A Generalization of an Equation of Chandrasekhar," *Differencial'nye Uravnenija*, Vol. 2, 1966, pp. 1264–1270; *Math. Reviews*, Vol. 34, 1967, p. 576.

R. F. Miller, "On a Nonlinear Integral Equation Occurring in Diffraction Theory," *Proc. Cambridge Philos. Soc.*, Vol. 62, 1966, pp. 249–261.

§16.24. See

R. Bellman, H. Kagiwada, R. Kalaba, and M. Prestrud, *Invariant Imbedding and Time-dependent Processes*, Amer. Elsevier, New York, 1964.

§16.25. See the application to the solution of the two-dimensional heat equation in

E. Angel and R. Bellman, *Dynamic Programming and Partial Differential Equations*, Academic Press, New York, 1972.

§16.26. See

R. Bellman, J. Casti, and B. Kashef, *Differential Quadrature: A Technique for the Rapid Solution of Nonlinear Partial Differential Equations*, (to appear).

R. Bellman, B. Kashef, and R. Vasudevan, *Differential Quadrature and Identification Techniques*, (to appear).

This is part of a general problem of replacing one operation by a linear combination of other operations. Very little has been done in this area. See also

R. Bellman, "Differential Quadrature and Long-Term Integration," *J. Math. Anal. Appl.*, Vol. 34, 1971, pp. 235–238.

Some early applications of this technique may be found in

R. Bellman, H. Kagiwada, and R. Kalaba, "On a New Approach to the Numerical Solution of a Class of Partial Differential Integral Equations of Transport Theory," *Proc. Nat. Acad. Sci. U.S.A.*, Vol. 54, 1965, pp. 1293–1296.

AUTHOR INDEX

N

O

P

R

S

T

U

V

SUBJECT INDEX